THE DATA MINDSET
PLAYBOOK

A Book About Data for People Who Don't Want to Read About Data

Gam Dias *&* Bernardo Crespo

Printed Worldwide
First Printing 2022
First Edition 2022

ISBN: 979-8-218-09285-6

10 9 8 7 6 5 4 3 2 1

FROM BERNARDO:

To my beloved Charo, Mateo, papá and mamá, tita Lola, Antonio, Elisa, and their families.

FROM GAM:

To K, A & T, who taught me what I needed to know, and to A & F who are still teaching me.

TABLE OF CONTENTS

FOREWORD

I am a physicist by training, and I completed my master's thesis in high energy physics at CERN in Geneva, the Conseil Européen pour la Recherche Nucléaire studying what can be learned from the traces sub-atomic particles leave in detectors. I then did my PhD at Stanford in neural networks and worked at Goldman Sachs on applying neural nets to predict and understand the traces Wall Street traders leave in the markets. Subsequently I became the Chief Scientist at Amazon in Seattle to find out what can be learned by studying the traces customers leave on the web. In each of these roles I immersed myself in the analysis of these traces, of particles, money, and people. Today I help businesses and governments look into the future, to help them imagine what the trails will be that will be generated and create strategies to unlock the tremendous value in those data.

I first met Gam Dias, the author of this book in 2010 at my birthday party in San Francisco. He had just co-founded an ecommerce company. I had been fascinated by his experiments of how digital personas fabricated across social media sites can act as a real humans and be perceived as real humans.

In my own book, Data for the People I describe it in the following way, "The recruiter had noticed Rebecca's profile on LinkedIn; she seemed to be very young marketing professional on the rise, someone who gets kudos from her colleagues, including a few from her days as an intern at a well-respected Silicon Valley firm. It was an awkward situation, however, because Rebecca is not a real person. She's the invention of our common friend, who decided to see how hard it would be to create and maintain a fake personality online".

Being able to see a handful of mutual confirmations and interactions is often sufficient evidence of a person's humanity, as Rebecca's profile can attest. So, I was wondering what makes a fake person with a real Facebook or LinkedIn profile appear so authentic, more authentic than a stranger with no profile at all?"

Gam became a core member of the Social Data Lab at Stanford and UC Berkeley. Since then, he has joined me on stage at international conferences as I greatly enjoy working with him in the data workshops we are doing together and have done together for many clients.

Gam created the data innovation track at IE Business School's digital transformation program. With Bernardo they are bringing the data mindset, the data ideology if you will to the next generation of executives. With their thoughtful observations on the use and misuse of data and business problems they bring their experience from the boardroom to the classroom. When I read their book I was very happy to see many of the ideas we had created in our conversations of the past decade written up quite clearly.

There's a famous saying about a bank robber who responded to the question, 'Why do you rob banks?' with 'That's where the money is.' When I look back over the last fifty years of technology adoption I think innovation has moved from universities to companies. Why? Because that's where the data is. These data are created by billions of mobile devices in the hands of the world's netizens whether they know it or not, whether they want it or not.

Previous hype cycles included digital twins and metaverses, now there is quantum computing, large language models like ChatGPT force us to rethink the goals of education and are actually challenging what it means to be human. On the physical side there will be nano-machines operating inside our very bodies, and much more.

When I was asked a few years ago to give a talk to the general assembly of the United Nations in New York about data I picked as the title 'Data is the new oil'. When Gam learned about this he made a very good remark that I incorporated "Not new oil, but new soil". He's right, there's no fear of us running out of data. Thus, the game is not about scarcity, but about abundance. And there, different rules apply.

The Data Mindset Playbook provides a playful perspective on the powerful world of data. In my talks and workshops, I emphasize the importance of starting with good questions, not with the data. And I make the point that it's not about actionable outcomes, but about actual actions.

No doubt that many of today's technology darlings will be superseded sooner than we think but the stories that led to them and perhaps most importantly our human curiosity is what hopefully persists I invite you to read this collection of stories, to take the time to reflect upon them and then to make this data mindset your own and then to apply it.

Data is free. Computation is free. What is next?

Dr. Andreas Weigend, San Francisco, 2022

Author, Data for the People

PREFACE

Not long-ago businesses tasked their IT Departments with finding answers. Answers become more plentiful as databases and reporting and analytics software proliferated. As everybody was able to find answers, they stopped bringing us competitive advantage. What enables today's companies and individuals to stand out from the crowd is asking better questions.

The most impactful questions are the most naïve, like "Why is it so difficult to get a cab when I need one?" That simple question, when combined with an understanding of data, how to find it and analyze it, led to the rideshare concept. Its business model challenged the legacy city transportation norms on a worldwide scale and demonstrated how the right questions can change industries.

For over a decade we have accompanied leaders of organizations on their transformation journeys. From global giants to worthy non-profits, from agile upstarts to some of the oldest institutions, we have helped them develop and execute their data strategies. Our interventions have spanned teaching, workshops, and hands on consulting. Our aim was always to inspire and guide, their success came through their ability to collaborate and iterate.

We noticed that these organizations had something in common. Data had been the preserve of the technologists, which left the organization bereft of data literacy. Organizations had more data than they could ever imagine, but institutionally they lacked an imagination of how that data can be used to transform its business, their industry, and our world.

We have been at the data frontline as developers and business decision makers performing analyses and presenting insights. We have crafted the business cases that resulted in strategic investments in data. And along the way we have had the privilege to work, teach and present alongside some of the most innovative and influential thinkers with respect to data. Dr. Andreas Weigend, whose work has shaped how we use data and data uses us, urges us to question everything, "We are at the risk of accepting whatever the algorithms project on the wall as our entire reality". This type of critical thought is fundamental to the holistic data mindset that we need to advance ourselves.

When challenged with developing a data strategy, a web search returns whitepapers, templates, and framework diagrams galore. Diligently following their directions will result in a data strategy that will serve your organization well. But the danger is that your data strategy will be a clone of your competitors'. In class and with our clients we use pictures and stories to show how to think differently about data and find sources of differentiation.

In this book we've written the stories and drawn these pictures to fire your imagination. Stories that hint at a vision of what might be possible for your organization if you seek out new data, if you look at problems with a different lens, and if you challenge the assumptions that you have had all your life. The pictures and the stories work together to imprint a lesson that can be applied to your data routines and challenges alike.

Read it sequentially or open plays at random. Read it in one sitting, or over the next year. If you want structure and direction, we have included a study guide and a sample workshop agenda at the end of the book.

Whichever way you choose to read, we are pleased to be able to share our data mindset with you, and we hope you find it useful.

Madrid, August 2022

PLAY 0: A CONVERSATION ABOUT THE EMOTIONS OF SKIING

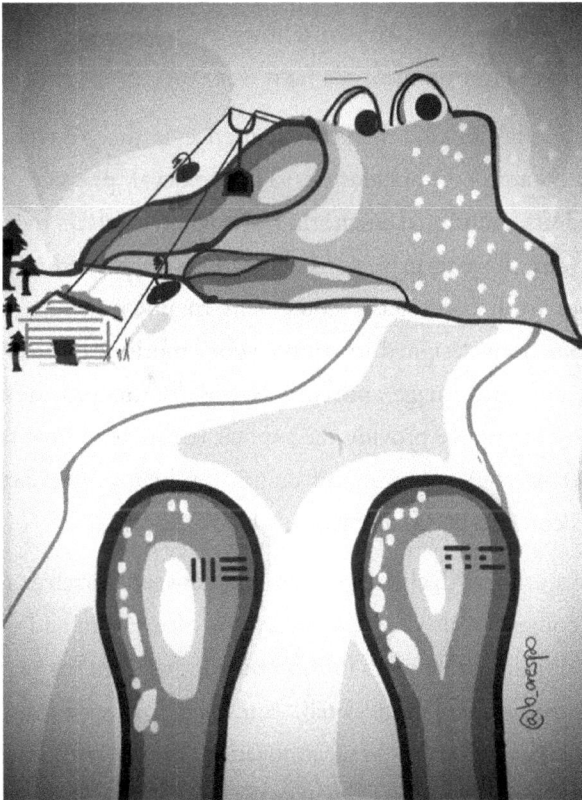

HOW DO STORIES HELP US COMMIT LEARNING TO MEMORY AND THEN PRACTICE? HOW DO CONVERSATIONS AND EMOTIONAL EXPERIENCES ACCELERATE THAT?

In our very first skiing lesson, we find ourselves standing with trepidation at the top of a slippery slope looking down. The teacher explains that we must lean into the slope to prevent us from falling down the slope, but to steer we have to trust our skis pointed down the gradient. The teacher is being perfectly clear, but for us as a novice fearful of taking an icy tumble, the instructions are counterintuitive and terrifying. However, there is a moment when the words and the action become one, mentally clicking into place and making bodily sense. When we look back with hindsight on that exact instant, we feel we'd known it from day one but never committed mind and body in unison until the light went on.

Such an emotionally charged experience is the difference between understanding and comprehension. The retrieval practice of a lesson learned, and its emotional anchoring are what catalyze the transfer of memories from short-term to long-term. Then with dedicated practice, learning goes from consciously competent to unconsciously competent. This book has been designed to trigger your emotions by beginning each play with a powerful image and curious anecdote to provide that mind-body learning before we provide the explicit lesson as to how to apply the data mindset to your strategic and tactical problems. The last stage, the unconscious competence as a practice, we leave to you.

We tell stories, with clients, in class and between ourselves as we debate data in the context of privacy, the metaverse, blockchain and the ethics of artificial intelligence. The plays in this book derive from the stories that we and our esteemed colleagues tell, and the conversations we have, conversations that have inspired innovation, conversations that shaped opinion and questioned long held assumptions.

Both of us are avid fans of the book The Cluetrain Manifesto. Early in the history of the World Wide Web, the book observed "A powerful global conversation has begun. Through the Internet, people are discovering and inventing new ways to share relevant knowledge with blinding speed. As a direct result, markets are getting smarter, ...and getting smarter faster than most companies". The Cluetrain Manifesto changed the paths of our careers and was foundational in writing this book. Whether you are a C-suite executive, a student, or just data curious, we really hope this book puts a spark to the data intelligence already within you, the data literacy and the data creativity that lies awaiting to catch fire.

Applying the Data Mindset:

When reading each play that follows — and it is no coincidence that there are fifty-two of them — think in advance about the future conversations that you will have with your colleagues, peers, and collaborators. There is your link between comprehending and transforming your reality. Tag every chapter with a sentence or a memory about the conversation you triggered after reading the play. This practice will help you to speak and communicate the language of data.

Choose a random week of the year and read back those plays that really inspired you and use this book to help others to get inspired. Soon after reading this book, we hope it will help others around you to start using the language of data, to write a new chapter of your own life, to change the lives of those you love, to create a better world that more of us can thrive in.

Reference

Levine, Locke, Searls & Weinberger. (1999) "The Cluetrain Manifesto: A foundational reflection about the present and the future of the Internet." https://www.cluetrain.com/

PLAY 1: SHOULD YOU SELL OR BUY YOUR CAR?

EVEN THOUGH WE MAY HAVE ENOUGH DATA TO ANSWER A QUESTION, WE MAY NOT LIKE THE ANSWER WE GET. LOOKING AT THE SAME INFORMATION FROM AN ALTERNATIVE PERSPECTIVE CAN HELP OVERCOME OUR BIASES.

As you see fuel prices rise, you also notice the maintenance costs for your old car steadily rising. You're torn between selling it or keeping it. You argue in favor of the increasing expense, that it's more convenient than taking public transport, taxis or rideshare.

How can you resolve your dilemma with a decision that you're happy with? What if you imagine that you have already sold your car and are considering buying it back for the same price, now would the pros of convenience outweigh the cons of increasing cost?

Tewari Analysis from the game of 'Go' suggests creating an alternative but logically equivalent scenario to resolve an argument or make decisions. Changing the scenario can help overcome status quo bias, something that is very useful when looking at a process from a data perspective.

Applying the Data Mindset:

Is there a problem that can be solved with additional data, data that requires expensive infrastructure to acquire, prepare and maintain? Can you justify the cost to solve the problem at hand?

To reverse the scenario, focus on the data and imagine how your existing business could be improved with the new data? What about if the new infrastructure can provision even more data? Can you provide a better customer experience or reduce costs with your suppliers?

Tewari Analysis would ask if you don't make the investment, what do you stand to lose? Plan a set of scenarios in detail looking closely at the costs and benefits to construct a robust ROI case for new infrastructure. When a CFO can compare a data infrastructure project with a proposal to add a new building or piece of machinery, then the data project will stand a better chance.

References

Author Unknown. "Tewari Analysis in Real Life"
http://fuseki.net/home/Tewari-Analysis-in-Real-Life.html

PLAY 2: WHAT DOES YOUR ALARM CLOCK SAY ABOUT YOU?

VALUABLE DATA SOURCES CAN BE HIDDEN UNDER YOUR VERY NOSE. BUT TO SEE THEM, YOU MAY NEED TO THINK LATERALLY.

Some people set an alarm clock and when they hear the alarm, they get out of bed immediately. Other people with natural, internal clocks don't even need an alarm clock. Others might hit the snooze button three or four times before actually getting out of bed. Which one are you?

Let's look at the data. There are people who set an alarm and people who don't. For the alarm-setters, look at the distribution of how many times people hit the snooze button and look at snooze button usage by day of week and by time of day. These are the behavioral cohorts of people who wake in certain ways each day at various times.

If you have other data about the population, and in that data can we find predictors of other traits? Do they make amendments to their meeting calendar, postponing and rescheduling meetings? Do people who repeatedly hit the snooze button shop differently online? Will they leave things in the shopping cart over multiple visits? If there are correlations, then how can we use them?

Where can we collect snooze data? Today's alarm clock is most likely a smartphone, a tool very adept at collecting data. If your mobile app included a clock with a snooze button, you could capture that data and relate it to other behaviors. If repeatedly silencing notifications marks the same behavior as an alarm snooze, then an email app provides similar data.

Applying the Data Mindset:

You are looking to understand customer or employee behavior. Is there something that people do that might reveal a trait that is important to you? What creative ways are there to collect the data that will reveal those traits?

PLAY 3: THE DATA ALCHEMY OF WEB SEARCH

WEB SEARCH HISTORY IS A RICH MINE OF DATA ABOUT PEOPLES'
INTENTIONS. HOW CAN THIS EXHAUST DATA BE TURNED TO GOLD?

The World Wide Web is growing all the time and navigating it becomes more complex every second. The most popular web search engines allow users to search the web free of charge. Free search has changed our lives.

We search for shoes and plumbers, we look for music, we check our own flu symptoms and learn how to make sourdough bread. In doing so, we leave behind us a data trail of the search terms we used and what we clicked on. This is the data exhaust of the search process, data that does not seem immediately valuable.

The search engine providers analyze search and click history to continually improve search results by ranking the relative popularity of each site for any search term.

They then flip the focus of the data analysis from site to user. Looking at a user's search history over time, they gain deep insight into what each user is thinking and doing. A user remodeling their home will follow a sequence of searches from design ideas to furnishings, to contractors, to materials. Preparing for the birth of a baby, the user reads reviews of hospitals, searches for advice, baby clothes, nursery equipment. The sequence is so precise that the birthdate can be estimated to sufficient accuracy. Intent, timing, and preferences are revealed.

Search engine advertising is a technique to monetize that search data using real time auctions of advertising space based on the search term entered. To feed the auction engines, search companies continually study and analyze our preferences, habits, and vices. The search advertising model stimulates content creation, content that can be searched feeding search advertising.

Is there a virtuous cycle that will allow your data project to grow exponentially?

Applying the Data Mindset:

What is the data exhaust being created out of your business or operation? When you collect and analyze that data, what basic insights do you find?

What might you discover if you flip the perspective of your analysis? How can you use this insight to improve your own operations? Is there anyone else that might find the same or a related insight useful?

PLAY 4: HAVE YOU THOUGHT ABOUT MIXING COFFEE AND TEA?

BLENDING TWO VERY DIFFERENT CONSTRUCTS CAN REVEAL
SOMETHING NEW AND IMPROVED.

Coffee and tea are very common and much-loved drinks, but we rarely see them combined. Why not? YuenYeung, or Cofftea is three parts coffee and seven parts Hong Kong-style milk tea and is advertised as 'the best drink you're not drinking'.

Let's think of another unlikely combination – cars and trains. Traffic on fast roads behaves like a pressure wave, stopping and starting. One driver's tap on the brakes is amplified into a major slowdown. If each driver was to keep moving at a constant speed, and were able to change lanes without harsh braking, the frequency of chain-effect slowdowns might be reduced.

If cars were interlinked with a simple exchange of data, they could sense the car behind or in front braking. As each car joins the motorway traffic flow, it senses and matches the flow of traffic.

It won't be long before fully connected cars are able to drive together, operating as if they were trains. But even before cars are given full autonomy on the freeway, a simple mobile app could gamify each journey for more efficient driver behavior. Each time a car decelerates sharply the driver loses points, encouraging drivers to keep proper braking distance, which in turn reduces the pressure wave effect.

Applying the Data Mindset:

Look at two processes in your business value chain that are creating data but working independently?

Is there a connecting piece of data that will link the processes together, perhaps the common KPIs or actors?

How can this data exchange be leveraged to create something new out of a hybrid of two concepts?

References

Ewbank, Anne. "Yuenyeung, The perfect marriage of coffee and tea.", *Gastro Obscura,* https://www.atlasobscura.com/foods/yuenyeung

PLAY 5: WHAT HAPPENS WHEN YOU TRY ON SOMEONE ELSE'S HAT?

WHAT WOULD YOU SEE DIFFERENTLY THROUGH THE EYES OF SOMEONE WITH DIFFERENT SKILLS AND EXPERIENCE?

Everyone has their own function or industry expertise, like finance or transportation. What is yours? Put your perspective to one side for a moment.

Think what you might see if you took another look at your business with another expert's eyes? If a human resources expert was to analyze supply chain data or a retailer was to look at healthcare data, what would they see?

Let's pick a pair of experts with different drivers. Firstly, the CFO whose concerns are predictability, cashflow and risk. And secondly a currency trader who is thinking about how fast they can spot an advantage before the opportunity disappears.

Every CFO is concerned about managing expenses. They run reports on submitted expenses by department and employee so they can find the top-spenders and identify overly high expenditure categories. They can evaluate cash advances, possible double claims and maybe who was taking too long to submit their expense reports.

Now how would the currency trader look at expenses data? Currency traders look at relative movements between currency pairs and forecast trends to exploit an exchange rate. A currency trader analyzing submitted expenses data from airlines and hotel chains would look at how expenditures with each vendor are trending over time. These insights could be used to negotiate volume discounts or referral agreements.

Applying the Data Mindset:

Take the business or data problem you are looking at and play the role of an expert from a different domain. Ask the questions that they would ask and see how these may give you a different perspective, perhaps creating new opportunities for increasing revenues or reducing costs.

PLAY 6: VAN HALEN'S NO BROWN M&MS CLAUSE

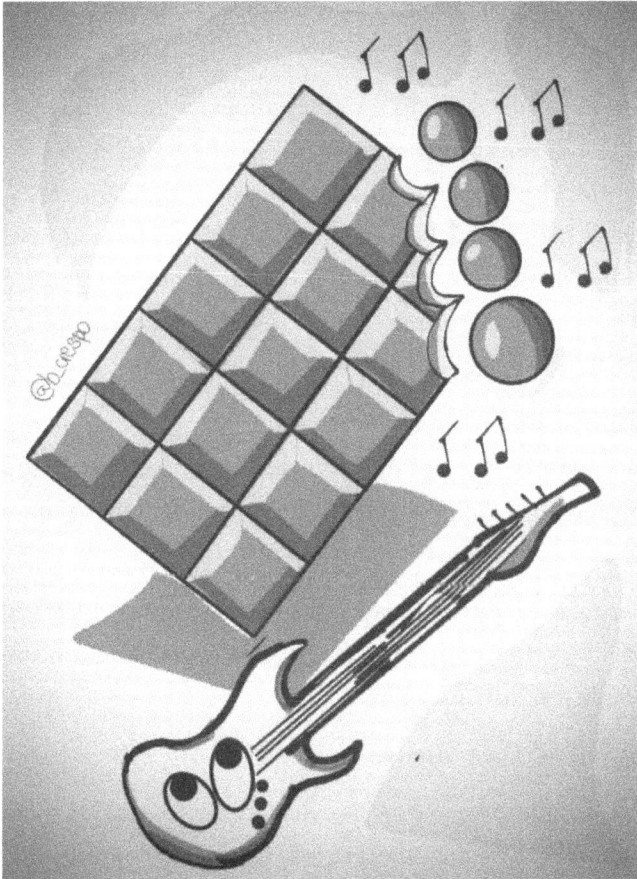

WE MAY DESIGN COMPLEX PERFORMANCE MANAGEMENT SYSTEMS, BUT THERE ARE EXCEPTIONS THAT STILL FALL THROUGH THE CRACKS. HOW CAN WE BUILD THE CHECKS AND BALANCES INTO THE PROCESS ITSELF?

The band Van Halen was famous for a rider in tour contracts that specified a large bowl of M&Ms in the band's changing room, but with all brown M&Ms removed. On the surface it seems like rock star excess, but in fact the reality is quite different. It was a trigger clause.

Van Halen was the first band to take a new and complex lightshow on tour, containing equipment that required special facilities and accommodations to ensure safety.

Promoters were notorious for not reading the contracts - bear in mind that the riders to contracts were often the size of phone books. Van Halen inserted the M&Ms clause into the very middle of the contract. "There will be 12 ampere high voltage sockets placed at 15 feet, not to exceed…" etc. And just after that important safety instruction, they would place another clause: "There will be no brown M&Ms found in the backstage area, or the promoter will forfeit the entire show at full price."

If the band arrived for a gig and found brown M&Ms on the catering table, that was a sure sign that the promoter had not read the contract. This warning would highlight potential safety issues and prompt a thorough line check.

Self-regulating systems identify problems early and trigger actions that will hopefully isolate and contain the issue. Where data is concerned, such self-regulation increases scalability.

Applying the Data Mindset:

When designing a new process, what are the constraints applied to the process, or indicators that the process is under-performing? Are there specific risks or compliance obligations that need to be considered?

Think of what additional data collection or information-sharing component can be easily added that would allow the process to be regulated?

References

"David Lee Roth tells the story behind the 'no brown M&Ms' legend", ImBigOnReddit channel on YouTube, http://youtu.be/_IxqdAgNJck

PLAY 7: WHAT IS THE MINIMUM VIABLE INSIGHT?

WHAT HAPPENS WHEN A TEAM IS DESPERATE TO FIND A GREAT ANALYSIS INSIGHT. THE EFFECTS OF THE SUNK COST FALLACY AND OUTCOME BASED THINKING.

A manufacturer of connected devices designed them to also send back performance data to monitor the health of customer installations.

Typically, their customers would periodically upgrade or renew their device, or they would cancel the service and return the device. The sales operations analysts had a hunch that data from installed devices can be used to predict what each customer would do next, renew, upgrade, or cancel.

The team decided to set up a project to carry out the analysis. Based on their large estimate of potential benefits, a significant budget was approved to acquire and prepare data and to extend existing infrastructure for processing and storage.

It took 6 months to prepare the data and perform the analysis. However, they were still unable to make an accurate prediction. They extended the project by another 6 months to look for better data. They found a weak correlation between poorly performing devices that required support and customer renewals. This correlation was just enough to present to their sponsors who followed the recommendation that a new installed base sales team was hired.

A year later, the installed base sales department was disbanded after having little effect on renewals and the analysts had moved on to other projects. The company knew it had wasted time and money although no formal analysis of the decision making that conceived and approved the project. The team committed time and effort and put their reputation on the line - which introduced a cognitive bias known as 'the sunk cost fallacy'. The company made an irrational decision because of factors that no longer impact the current alternatives.

There was a bigger mistake. The company analyzed the failed outcome, terminated the investment, ending the project, and they disbanded the teams.

They did not take the time to understand the system of decisions that led to the mistake. This is the difference between outcome-based thinking and system-based thinking.

Applying the Data Mindset:

When a new project is proposed, take a Minimal Viable Insight approach where you do just enough analysis to prove that the next step will be feasible.

First build a solid hypothesis, test that with a small amount of data. On a positive result, increase the level of investment, then test again. You may discover that you have disproved your original hypothesis, but that's good science too. You can revise the hypothesis and test again with another small investment.

Remember the data will tell you something, although it's not what you might want to hear. Start small and iterate. Baby Steps.

References

Are we likely to continue with an investment even if it would be rational to give it up?' The DecisionLab, https://thedecisionlab.com/biases/the-sunk-cost-fallacy

Bahcall, Safi. 'The Decision-Making Secret Shared by Pixar and a World Chess Champion' Marker Medium, December 16, 2019, https://marker.medium.com/the-decision-making-secret-shared-by-pixar-and-a-world-chess-champion-538c69015ee7

PLAY 8: GIVE DATA TO GET DATA

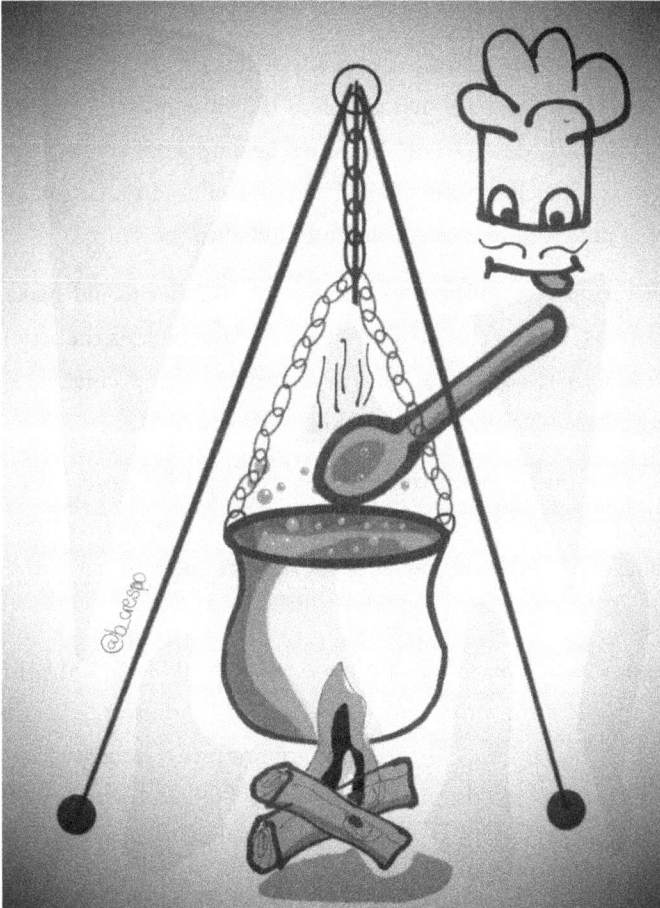

HOW CAN YOU HOPE TO BEGIN AN ANALYSIS PROJECT WHEN YOU
HAVE ABSOLUTELY NO DATA? THE FOLKTALE OF STONE SOUP
PROVIDES A VALUABLE APPROACH.

A weary traveler stopped for the night in a village. She asked where she might get an evening meal. The villagers, barely surviving themselves, were unwilling to offer the traveler anything to eat.

Undeterred, the traveler produced from her pocket a smooth rock, she asked for a pot of water and wood to make a fire. She placed the rock into the water and began to stir. "I shall make stone soup", she announced.

After some time, she tasted the soup, licked her lips, and said, "this is a fine soup, but it lacks a touch of salt". One villager came forward with a salt cellar, she added some salt and tasted again. "Thank you", replied the traveler, "what excellent soup, but if only I could add a carrot". Another villager was able to find a carrot that was added to the soup.

"This soup is exquisite, but a couple of potatoes would make it even more delicious" Again, the villagers were able to oblige. As the aroma of the cooking soup pervaded the village, an onion, some chicken bones, a cabbage, and some sausage were all donated and added.

"This is truly the most rich and hearty soup I have ever tasted, please bring your bowls and spoons and share it with me" announced the traveler to the villagers. And with that, they all had the most filling and delicious meal.

Airline loyalty programs are a stone soup of data. On joining the program flyers tell the airline whenever they buy and use a ticket. In return flyers can see their mileage history and earn rewards. The airline is now able to understand how a customer searches for flights, what they pay, how much in advance they buy and where they travel to. The intelligence gained is worth far more than the rewards offered.

Applying the Data Mindset:

Are you short of important data with no obvious way of acquiring it? Do you have some exhaust data with which you can create a valuable proposition for the customer? One that could be used to collect more data.

Look at the data assets you already have. Think about the missing data you need. Imagine a pipe that is carrying data into your business from your customers or partners. If data can flow one way in that pipe, it can flow in both directions just as easily.

What data can you first share with your customers or partners that will encourage them to share some data of value with you?

References

'Stone Soup, An Old English Parable', BeliefNet
https://www.beliefnet.com/faiths/pagan-and-earth-based/2002/01/stone-soup-an-old-english-parable.aspx

Link, Rachel. 'Stone Soup Rocks in Remote Oaxaca' National Geographic, October 20, 215

https://www.nationalgeographic.com/culture/article/follow-the-path-of-the-real-stone-soup-to-remote-oaxaca

PLAY 9: YOU HAVE BETTER DATA THAN THE BEST SOCIAL MEDIA SITES

YOUR BUSINESS MAY ALREADY HAVE BETTER QUALITY DATA THAN ANY SOCIAL MEDIA PLATFORM. YOU SHOULD NOT BE SURPRISED. BUT BE SURE TO RESPECT PRIVACY AS YOU LOOK TO CREATE NEW VALUE.

You may have many social followers but will have only physically met a small percentage. Your social community friends could include a number of people that you don't even know. In fact, much of the data presented in social media is untrue and even actual users are fake. Birthday, hometown, employers - anything and everything.

Professional social networks are a more truthful space just because they act as an extension of our professional environments. But what would happen if a person added a few false jobs in there – who would check? Likely nobody unless you listed yourself president of a country and even then, you'd probably get away with it.

When it comes to analyzing social network data and using it for propensity studies, risk analysis or behavioral profiling – remember that social data is likely to be fictional. You can use other data to validate the data you are studying, for example does each person's social network fit the profile expected? Does their direct message history match the profile of their published posts? If home addresses are provided, can street photography validate the building as residential?

In your own organization if you have actual transactions with real people – perhaps with actual physical addresses or credit card purchases - this data is authenticated and therefore more valuable. Direct communications between two parties over a period are very real and tangible data points that can prove a relationship.

This type of data can be more valuable inside your organization than any social media data, but releasing it creates risk. In 2007 a team took publicly released anonymized viewing data from a streaming movie network and combined it with movie reviews online by self-identified users. The team was able to identify the individuals from the released data, potentially putting the users in harm's way.

Applying the Data Mindset:

Is your organization collecting valuable personal data that could be used to create value? Consider how that data could be used and ensure that as you collect it you are transparent with the data subjects as to how it will be used. Be sure to gain their explicit permission.

With permission, that data could be used to create new customer value propositions. If the value to your customers is high enough and you continue to respect their privacy, they will share even richer data with you.

References

Weigend, Andreas. 'Data for the People, - Chapter 3' https://weigend.com/en

Rosenbaum, Eric. 'Forget IPO value, what's Twitter's data worth?' CNBC October 12, 2013 http://www.cnbc.com/id/101103596#!

Maldonado, Sergio. 'Zero-Party Data vs. Declared Data' Privacy Cloud August 2, 2021, https://medium.com/privacycloud/zero-party-data-vs-declared-data-3cd5b2913667

Schneier, Bruce Why 'Anonymous' data sometimes isn't December 12, 2007, Wired Magazine https://www.wired.com/2007/12/why-anonymous-data-sometimes-isnt/

PLAY 10: THE RACING FORMBOOK

IF YOU HAD TO CREATE A FANTASY PROJECT TEAM, WHAT METRICS WOULD YOU MEASURE? WHERE CAN YOU DRAW DATA TO BE ABLE TO ASSEMBLE A WINNING TEAM?

Consulting companies are masters of managing resources. Balancing the number of consultants against the number of client projects is essential to profitability. On the other hand, clients want to get the best consultants for their particular project.

A lot of time and energy goes into selecting the best consulting team to work on a client project. This decision process is where we can find an opportunity to use data.

Choosing a consultant is like placing a bet on a horse. In the sport of horse racing, gamblers have a formbook that shows the track record for each horse in a race - a history of recent performance, which other horses were running and what position the horse finally finished. The formbook also shows the racecourse, jockey, type of race, weather, and ground conditions it prefers and weight information. This detailed data allows the gambler to make a better decision.

Keeping a formbook for each consultant assigned to a project captures project history, other consultants on the team and the success rate of projects in terms of objectives, costs, and schedules just like a racehorse. The consultant formbook would include which combinations of consultants worked well together and created the best outcomes.

There is a danger that the formbook becomes biased, favoring consultants in a narrow group, from certain schools, backgrounds, and genders, if that has been the historical pattern. A homogenous portfolio of consultants will reduce diversity of thought and the ability to truly innovate.

Applying the Data Mindset:

Is there a resource allocation problem that you are dealing with? How can you find the best available resource for the job? Can you leverage historical data to make better decisions? How can you mitigate potential biases in the hiring system to increase diversity of thought into the team?

What processes can you set up mid-cycle to ensure that historical data is collected and available for future use? Do you have permission from all parties to collect and use that data? What benefits would you explain to the resources that would encourage them to share their data?

PLAY 11: IF ONLY I'D KNOWN

BUSINESS DECISION-MAKERS OFTEN COMPLAIN THAT THEY ARE "DROWNING IN DATA BUT HAVE NO INFORMATION". CAN WE USE 20/20 HINDSIGHT TO PREVENT THIS?

They say hindsight is 20/20. Have you ever made a decision, and later wished you'd done something else?

Traveled for a business meeting, then realized both parties would be at the same conference the next week? Purchased a box of paperclips then discovered ten boxes in another department?

Data should help you analyze the right data at the right moment when you're about to make a decision, so you never have to say, "if only I'd known".

As I am scheduling that meeting, can I see the other party's future travel plans (if they allow it)? As I am ordering a product, let me see office inventory and location of that product or substitutes.

In large organizations there is often a tremendous backlog of decision makers asking for reports. To satisfy as many users as possible, a lot of reports are generated. The most time-consuming activity is normally finding and preparing the data and formatting the report. Which can result in a voluminous report with a rich set of variable parameters, but that report provides great answers to the wrong questions.

Great business intelligence allows for sufficient time to truly understand the decisions being supported. Sometimes, even the basis of the decision making is brought into question to produce better insights. Role-play the decision and its impacts to see what 20/20 hindsight might have to offer.

Applying the Data Mindset:

Examine a business process in your organization and list the decisions at every stage. What information is used to support that decision?

What are the undesirable or suboptimal outcomes of that decision? Is there another piece of information that would guide a better decision. How would you get the data, process it, and present it to the decision maker before they made their decisions?

PLAY 12: WHO IS TALKING TO WHOM?

Every activity in an organization or between organizations is generating some sort of data. When we look deeply, we start to notice the data and we can imagine how we could analyze it

There are plenty of online conferencing services available, and they are relatively easy to use. A host schedules a meeting and invites other participants. Everyone dials in at the scheduled time, has their call and then they all drop off. Making video and audio conference calls run smoothly over multiple, often patchy, internet connections is a technical challenge that has largely been solved. In the background valuable data is being generated.

First, the host provides some details to set up the call, the host's email and call title and timing. During the call we can see how many people participated and for how long, the actual call duration, which people showed up, typed interactions from the chat, who speaks the most and the least. Recurring calls will reveal patterns such as how many participants, time of day, locations, who are the leaders and who does not engage.

What can be done with that data? During the call, the voice can be accurately transcribed to text. The conversations can be analyzed in various ways, in real time to provide translations and relevant information either from inside the firewall or from the web. Offline the texts can be analyzed to see what topics were discussed with what intensity.

Can the conference call company send a management report to its clients noting organizational behaviors? Might an HR department find this useful as it tries to create a smarter workplace? Can we map the interactions between departments or functional groups? Added to email traffic, we can see how effectively an organization communicates. Can the level of conference activities between company email addresses predict business deals or other patterns between organizations?

Obviously, this data cannot be used without the clients' or individuals' permissions but examining the data underpinning something as routine as a conference call allows us to see what is possible.

Applying the Data Mindset:

Think about a business process you are working with. Who are the actors, what actions do they perform individually and together? What variations in behavior are there between different people, and at different times? What data is being generated and how much is collected?

Look deeply within the business process you are working with to build an extra detailed entity model beyond that required to support the process. The data you collect about certain entities can help you see patterns and trends that can predict outcomes in other areas of your business.

PLAY 13: THE LIFE OF A SINGLE AIR MILE

OFTEN OUR EXPERIENCE GUIDES HOW WE THINK ABOUT
POPULATIONS, BUT THE UNDERLYING DATA THEY GENERATE CAN
PROVIDE AN ALTERNATIVE PERSPECTIVE TO FIND NEW WAYS OF
CREATING VALUE.

Marketers define customer segments, cohorts of similar customers, to understand relationships and interactions over time. The distinct needs of each segment can be used to create more useful products and services, and to customize and target messaging.

Airlines have developed some of the most sophisticated customer loyalty programs to analyze the flying patterns of millions of members. Airlines segment flyers by type of traveler, for example motivation (business travelers, leisure), age, nationality etc. For example, business travelers will pay more for a lay-flat bed because it will allow them to arrive fully rested and ready to work. Yet the segments are usually defined by marketers top down based on their experience and intuition.

An alternative way of segmenting would look deeply into the data. What if we analyzed the lifecycle of each individual air mile flown by a traveler? What type of flight the mile was earned on, how long did it sit in the member's account and then how it was spent.

Two business travelers might have exactly the same number of miles in their accounts, but their earning and usage behaviors are distinctly different. One might earn their miles on a small number of intercontinental trips and spend them immediately on family vacations. Another might fly many national flights and save their miles for years before exchanging them for upgrades. Clearly these two travelers can have very different product and offer profiles.

Studying the entire member population will provide insights into broad flier patterns across entire segments.

Applying the Data Mindset:

Consider your business process as a set of unitary transactions that each customer or stakeholder performs. Most processes have multiple steps, but underneath, there may be a common micro-transaction that is replicated for all steps. That micro-transaction can be used as the base aggregate to reveal good and bad performance.

PLAY 14: CINDERELLA, YOU SHALL GO TO THE BALL

EVERY BUSINESS HAS A SET OF NON-GLAMOROUS FUNCTIONS THAT PROCESS HIGH VOLUMES OF TRANSACTIONS. THOSE OFF-VALUE CHAIN ACTIVITIES GENERATE A LOT OF DATA WITH A HIGH POTENTIAL VALUE.

A value chain is a set of business functions needed to deliver a valuable product or service to customers. For example, the retail value chain would include buying, supply chain, warehousing, distribution, stores, and customer service. A dental practice has a procurement function for equipment and consumables, patient intake and scheduling, dental operations, and billing.

Value chains are built on the organizational process view of a product or service organization as a system, made up of subsystems each with its own inputs, transformation processes and outputs. Value chain execution impacts costs and affects profits.

Organizations use data to optimize their unique value chain. An automobile manufacturer focuses on manufacturing and supply chain data to increase responsiveness and lower costs. Drug development and testing data is key to success for pharmaceutical companies. Mobile telephone providers will analyze call data extensively.

According to Porter's model, each organization needs support functions: finance, legal, procurement, and HR. These functions are often outsourced, and efficiency becomes the outsourcer's problem.

Data created by the back-office functions, the "Cinderella" organizations, has high value. For example, invoice processing, process logs and payroll and timesheet data can be analyzed to find new sources of benefit. Furthermore, dark data such as old customer support tickets or follow up emails can be used to spot patterns, trends, and anomalies.

These back-office functions outside the value chain are often neglected in analysis due a low unit value, but their high frequency means they have a large impact. For example, in an accounts payable department, if all suppliers are treated the same, the administration required to validate and approve each is high.

To fast-track reliable suppliers saves a huge effort and time, freeing experienced resources to deal with new or vendors more prone to invoicing errors.

Applying the Data Mindset:

List out the support functions in your organization. Then speak to someone in each function and ask if they need any new reports or analytics. Think about the data that is being collected or being generated by those groups and how it is used. Whether you carry out simple analysis or combine currently siloed data sources, you might find opportunities to create a lot of value.

References

The value chain was first popularized by Michael Porter in his 1985 best-seller, Competitive Advantage: Creating and Sustaining Superior Performance. http://resource.1st.ir/PortalImageDb/ScientificContent/182225f9-188a-4f24-ad2a-05b1d8944668/Competitive%20Advantage.pdf

Rindler, Eric The value chain June 2018 Inside Dentistry https://www.aegisdentalnetwork.com/id/2018/06/the-value-chain

The Rise of Dark Data and What It Means to Accounts Payable. Kefron https://kefron.com/uk/2016/07/the-rise-of-dark-data-and-what-it-means-to-accounts-payable/

PLAY 15: USING DATA TO DEPLOY FLOOD BARRIERS

WE OFTEN MONITOR METRICS ON A COST OR A BUSINESS ISSUE.
BUT MAYBE THE METRIC WE ARE WATCHING WILL SOLVE OR
REMEDY THE PROBLEM ITSELF.

An insurance company writes policies to cover peoples' homes against damage to buildings and contents. To assess risk, they look at how the building is constructed, how far it is from a fire hydrant, local weather conditions, and other characteristics of the building and area.

Where the building is close to water – a river, lake, or sea - they also look at the flood patterns in the area. They can do this using a flood map that calculates the probability of the building being flooded after a certain amount of rainfall.

For buildings at greatest risk of flooding after a small amount of rain, they send the owners of the building a device that minimizes flood damage - a waterproof fabric barrier that can be quickly unpacked and wrapped around the house. A vertical part to cover the walls and a horizontal part extending onto the ground. As the water rises, the part on the floor is held down by the weight of the water and the part on the walls forming a barrier. When the weather forecast predicts heavy rain, homeowners get an alert to deploy their barrier.

For homes that are flood prone, this system uses the flood report data to save the property and all the contents from the huge damage of flood waters. For the insurance company, this vastly reduces the cost of the claims pay-out.

Applying the Data Mindset:

What buckets of cost to your business operations are being reported against? Pick one of the cost items. What actions can be taken to reduce the average cost? What information would be useful in taking that action or to obtain the required outcome? Where can you reliably obtain that information?

References

'What is a flood map?' https://www.floodsmart.gov/all-about-flood-maps

PLAY 16: THE CAR, THE DRIVER, AND THE ROAD

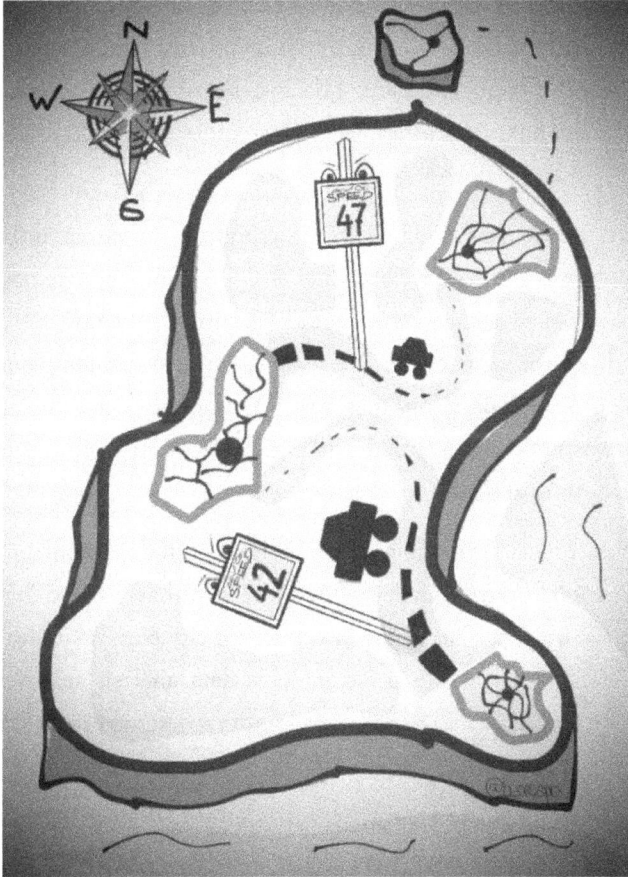

CONNECTED DEVICES ARE PROLIFERATING, AND WITH THEM THE STREAMS OF DATA THEY GENERATE. ANALYZING THAT DATA CAN LEAD TO NEW PRODUCTS AND SERVICES THAT CAN BE TARGETED AT NEW MARKETS.

Cars are becoming smarter. Telematics equipment determines the precise speed and bearing of the car at any time, including the lead-up to an accident.

This is commercially valuable for auto manufacturers and insurers. The aggregated data can show driving patterns for any make and model of vehicle or by groups of drivers. Commonly believed stereotypes can be confirmed, for example drivers aged 17 to 25 years are at higher risk of accident between 10pm and 6am. This insight led to an insurance product designed for young drivers offering cheap premiums for miles driven during the day.

When the analysts switched the focal point from the car and driver to the road, they began to develop other insights from the telematics data. When they looked at a certain road, over time, they saw vehicles that regularly drove the same route. The analysts studied this sample of cars to determine average speed on that road. Using this data, the insurance company was now able to build a map of the country with average road speed at any time of day.

The insurance companies were able to build a commercial data product that showed road usage by hour, and how that changed during road construction projects. This created a service for city planners to forecast the effects of road construction projects. This data analysis spawned a new insurance proposition and a new data product designed for a new client sector.

Applying the Data Mindset:

What subject entities are the focus of your data collection? Now turn the focus on the other entities that are also involved in the process. What does the data look like when they are made the subject? Who else might be interested in that data? This is where commercial data products can be found.

References

IEEE Spectrum – The Radical Scope of Tesla's Data Hoard
https://spectrum.ieee.org/tesla-autopilot-data-scope#toggle-gdpr

National Association of Insurance Commissioners (NAIC)
https://content.naic.org/cipr-topics/telematicsusage-based-insurance

The Markup - Who is Collecting Data from Your Car?
https://themarkup.org/the-breakdown/2022/07/27/who-is-collecting-data-from-your-car

PLAY 17: PRINTER INK IS MORE EXPENSIVE THAN PERFUME

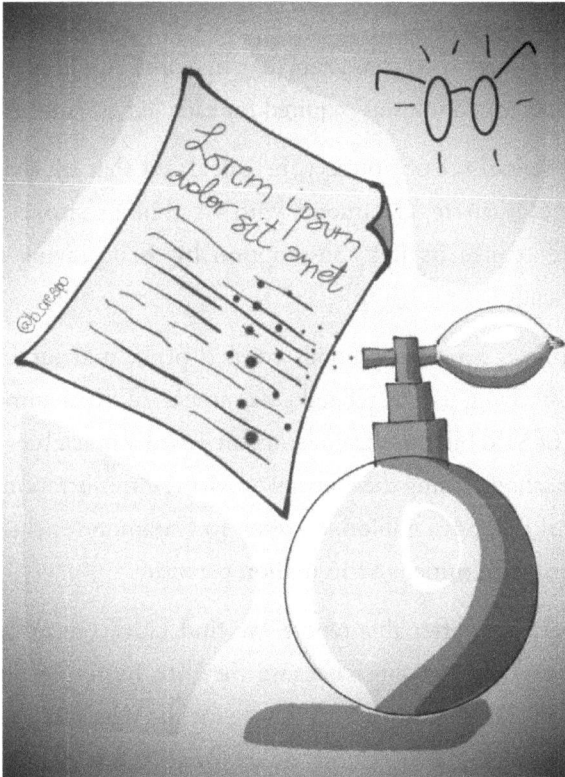

MOST PROBLEMS ARE NOT AS SIMPLE AS THEY SEEM ON THE SURFACE, HOWEVER, THEY CAN SOMETIMES BE SOLVED WITH A SMALL AND ELEGANT CHANGE. BUT ONLY IF YOU HAVE DISSECTED THE PROBLEM ENOUGH.

When a fourteen-year-old student started middle school, he received a lot more photocopied sheets than the previous year. His school's directive to save or recycle paper now left him confused.

He happened to read that printer ink was more expensive than designer perfume, ounce per ounce - possibly double. Inspired by this insight, he focused on finding ways to reduce the use of ink instead of saving paper.

The student collected random samples of teachers' handouts, and found the commonly used characters (e, t, a, o, and r). He then used a tool to measure how much ink was required for each letter across various fonts.

For his school science project, he calculated that by changing fonts from Times Roman to Garamond with its thinner strokes, his school district could reduce its ink consumption by 24%, saving as much as $21,000 annually.

The project won the school's science prize, and an invitation to propose a cost saving for the federal government. With an annual printing expenditure of $1.8 billion, the government posed a much bigger challenge than a single school. Using the General Services Administration's estimated annual cost of ink, $467 million, a switch to Garamond exclusively would save the federal government $136 million per year.

However, even after this report, we find Government texts are still printed in heavy serifed fonts because they are more legible for poorly sighted readers. Yet the valuable takeaway for the data infantry is the idea of solving a problem through looking literally at its fine print.

Another example is the trend of working from home. This has left vast swatches of office space unused during the week, workers coming in less than fifty percent of the time.

However, even with people working in the office 5-days per week, 8 hours per day, the building occupancy began below 25% of the 168 hours in a week. Solving this is not a space problem but a collaboration opportunity.

Applying the Data Mindset:

Have you been tasked with solving a complex business problem? First think of the most common solution - what is the math behind that solution. Now break down the entire situation by its variables and look at each one in turn, asking why each is that value. Which variables can you change and what might the effects be, both materially and psychologically?

References

Teen to government: Change your typeface, save millions https://edition.cnn.com/2014/03/27/living/student-money-saving-typeface-garamond-schools/index.html

APFill Ink and Toner Coverage Calculator https://download.cnet.com/APFill-Ink-and-Toner-Coverage-Calculator/3000-6675_4-10347540.html

Save $400M printing cost from font change? Not so fast… http://www.thomasphinney.com/2014/03/saving-400m-font/

The Metrics of Distributed Work by Knoll Workspace Research https://www.knoll.com/knollnewsdetail/the-metrics-of-distributed-work

PLAY 18: A TRAVELER WITHOUT A PASSPORT

BEING ABLE TO CONFIRM IDENTITY IN AN INCREASINGLY DIGITAL WORLD IS VITAL TO GOVERNMENT AND COMMERCIAL ORGANIZATIONS. THE PAPER DOCUMENTS WE CARRIED ARE RAPIDLY BEING REPLACED WITH ANALYSES OF PERSONAL DATA.

My jet-setting friend recently needed to return home urgently from her business trip. To complicate matters, her passport was stolen, and her country's consulate was unable to provide replacement travel documents in time. Her very high mileage frequent flier status meant the airline carried a lot of data on her. Using this data, as verified by the airline, she was able to pass through immigration control when she returned home.

The passport document evolved from a letter of recommendation between nations to the complex electronic record we have today. When coupled with a biometric scanner, it can be used to track movements in and out of a country.

As we see from the story even without a passport, a verified personal data trail is still a solid authenticator of identity. What non-personal data such as fingerprints, retina scans and DNA can be used to validate identity?

Phone call history and voiceprints, credit card receipts over a period of time will show frequencies of establishments and purchase characteristics. Social media interaction, GPS location of phone. All these data points can be used to confirm identity. Could you identify a close friend or family member by their travel patterns, and spending in restaurants and stores?

Identities historically proven by a document are now validated by a phone app or a biometric validation and as fast as the technologies are being developed, so are ways to defraud or spoof them. Governments are researching new ways to uniquely identify us in a digital age. Organizations can use this thinking to enhance their own security or authentication processes.

Applying the Data Mindset:

Your business may deal with individual members in a population – people, vehicles, products, or livestock. Is there a trail of data that can be used to provide a unique behavioral fingerprint that can uniquely identify a member? How else can this data trail be used legally and ethically?

References

The Fundamentals of Tech Transformation: Identity in a Digital Age https://institute.global/policy/fundamentals-tech-transformation-identity-digital-age

PLAY 19: IF YOU'RE GIVEN LEMONS, SHOULD YOU MAKE LEMONADE?

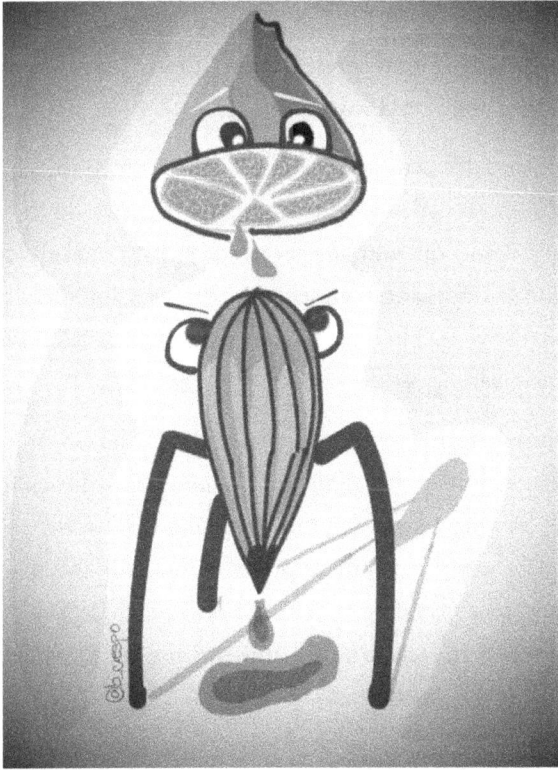

WHEN GIVEN A CHALLENGE AND PRESENTED UPFRONT WITH CONSTRAINTS, SOMETIMES THE CONSTRAINTS FILL OUR FIELD OF VISION SO OTHER POSSIBILITIES ARE BLOCKED OUT OF VIEW. WHAT HAPPENS WHEN WE ZOOM OUT?

An entrepreneurship class in a top Ivy League business school was given the challenge to maximize the return on investment of $5 in exactly 2 hour and 3 minutes

In teams that were each given $5, their goal was to make as much money as possible within 2 hours and then to give a 3-minute presentation.

Most of the groups took a highly logical path, they ran to buy materials for a makeshift car wash or lemonade stand, set up shop and began to trade. Groups that were risk takers bought a lottery ticket or wagered the money on a bet. In their 3-minute presentations these groups showed a small positive return for their hard work.

One group decided the $5 was a distraction. Instead, they brainstormed how to make as much money in the two hours starting with nothing. They came up with a clever scheme to make reservations at popular local restaurants and then selling the reservation times to those who wanted to skip the wait. This group managed to return $200 for their smart use of the time given.

The winning team however saw that both the $5 and the 2-hour period were constraints rather than assets. They stepped back to look at the bigger picture of what assets they had to make work for them. The most valuable resource was the 3-minute presentation time they had in front of the captive business school class. This group sold their 3-minute slot to a company interested in recruiting the students for $650. They were able to present an ROI of 12,900%.

Just because there is an asset available to you, it does not mean that you should use that asset and not consider any others. Zoom out so that other assets and possibilities come into the picture and focus on them.

Applying the Data Mindset:

Have you been challenged with a task, and given some resources and constraints? Look at each resource and determine if it alone is useful, it can be combined or even discarded. Look at your constraints, can they be legitimately broken?

Are there other resources that are available to you that you have not been specifically allocated and constraints or opportunities that were not mentioned in your scope. How can you use those to exceed the expected result.

References

The $5 Challenge How Stanford Students Turned $5 Into $650—in Just 2 Hours by Tina Seelig, Ph.D., Stanford School of Engineering. https://www.psychologytoday.com/us/blog/creativityrulz/200908/the-5-challenge

PLAY 20: WHAT WOULD YOU DO WITH A MILLION RESUMES?

COLLECTIONS OF THINGS HOLD VALUABLE DATA, ATTRIBUTE VALUES THAT CAN BE ANALYZED AND COMPARED. GROUPINGS OF THINGS, FILTERED SUBSETS ALSO PROVIDE OPPORTUNITIES FOR FURTHER ANALYSIS, SOMETIMES WITH SURPRISING INSIGHTS.

Are you on a professional social networking site, one where you share your resume details and connect with other professionals? As professionals on these sites, we freely share a lot of interesting data about ourselves.

As members of the site, we see resumes and connections. Recruiters can filter those resumes based on location, company, job title, school, or other searchable member attributes.

As data analysts looking at the data in aggregate, we see something much more interesting. We can create directories of all the employers and all the schools, all the job titles and all the qualifications.

Using job titles, we can distinguish junior versus senior jobs. Then we can determine the best colleges and degrees for different jobs. We can compare how fast people climb the career ladder. With this we can count how many other fast movers they are connected to, which qualifications fast movers typically gain, and how other professionals rate them.

The data sign-flip comes when we look at companies. Represented in a data model each organization is a collection of resumes where the current employer is that organization. As people change companies and update their resumes, we see an organization growing and another one shrinking. We can assess the quality of the team they recruit or lose by looking at employees' career velocities and reviews and recommendations. This could tell us the best companies to work for, or even companies that are likely to perform well in the future.

Applying the Data Mindset:

As you examine the data in your organization, be sure to build the data model of entities. Then consider if a transaction between entities might also be occurring. Change of status of an entity might provide a valuable clue to a transaction. If you aggregate those transactions, what patterns might emerge that could be useful?

PLAY 21: FINDING TRUST IN A DIGITAL WORLD?

WHAT ARE THE ELEMENTS THAT LEAD US TO TRUST SOMEONE IN THE REAL WORLD? HOW CAN THOSE FACTORS BE TRANSLATED WHEN OUR ONLY CONTACT IS DIGITAL?

Online consumers will mostly never meet anybody in the business that they buy from. Due to the high cost of voice contact when scaled, online businesses route customers to email or chat. For a customer, it's rare to hear anyone at the company's name.

Businesses that work digitally never meet their clients or suppliers. Some employees barely ever meet their bosses or co-workers. In digital business where social interactions are largely absent, how do we build trust?

Once we get to know a person by observing their reactions and experiencing their work, our confidence grows. Personal introductions, alumni groups and professional associations provide the assurance that we are dealing with someone 'similar' to ourselves, someone that should be trusted. Those connections also inform us of their published reputation and provide recourse for the future. In the same way they increase trust, exclusive groups reduce the universe of potential employees, partners, or customers, which stunts a healthy diversity.

Private members' clubs had a role in society, providing a physical venue for similarly minded people, with business networking usually implied. Such members' clubs do have digital equivalents. Online there are communities of stock traders, of specialists in every field, and forums where they interact and network.

The knowledge we sought to reassure us about a person boils down to information about them, their metadata. For online business connections, professional social media has made it easy to research a person, including recommendations from colleagues. Personal introductions including those via dating apps can be researched using social media. It's all data about each person, published, shared, corroborated, and verified by the community. We just must understand how these communities are created, members recruited, and maintained such that the network can thrive.

If you're looking to build trust in a digital world, sharing personal metadata accessible by other systems can be the fastest way.

Applying the Data Mindset:

What business is your organization engaged in and who needs to trust whom? Is there data that will provide that trust factor – history of transactions, relationships with other parties or shared association. How can your service be designed to leverage that trust? How can the user experience be created so that users share more of that trust-building data?

PLAY 22: WHO WILL BUY YOUR DEADHEAD MILES?

TO PROVIDE HIGH SERVICE LEVELS MOST SYSTEMS OPERATE
UNDER-CAPACITY, WASTING PERISHABLE RESOURCES. THE SHARING
ECONOMY IS A GIGANTIC DATA PRODUCT TO BROKER BOOKED AND
PAID FOR RESOURCES.

Peer-to-peer applications are famous for increasing the utilization of idle assets: Spare rooms in a home; Clothes in the closet; Equipment in a workshop; Times when you or your car are idle; Or just a seat in a vehicle driving somewhere.

The sharing economy apps provide mini marketplaces where those in need can find timely and available surplus. The best sharing apps made it easy for both parties to safely share data.

Deadhead miles are defined as any time a freight carrier is moving without revenue generating goods on board. Freight vehicles once they've made their delivery often return empty. Some suppliers made use of 'backhaul', using the returning vehicle to carry returns. However, high fuel prices catalyzed the move by shippers to track and utilize the empty capacity.

Freight marketplaces evolved to allow shippers to list empty truck routes that can be matched with shippers seeking to move their goods. There are emergent other data marketplaces on which new applications can be built, for example rail data and healthcare data.

Ideas like this emerge from taking a global perspective that highlights overcapacity and unmet demand for a perishable resource. Ocean going container vessels have a similar problem due to trade imbalances, this problem is still seeking a solution.

Applying the Data Mindset:

What resources are not fully utilized in your organization? What is the acceptable cost of those resources being non-operational? What real or artificial cost constraint can you find that will force you to increase utilization?

Who else might find value in the resources and what is that worth? How can you publish periods of downtime and make those resources available elsewhere? Are there established marketplaces for the data you are creating so that enterprising startups can create new and valuable services?

Conversely are there are resources that your organization needs but cannot produce enough of? Who might have a surplus that you can utilize? Is there a marketplace where those with surplus can be matched with those in demand?

References:

Benson, Ken Keeping Trucks Full, Coming and Going, April 22, 2010, New York Times https://www.nytimes.com/2010/04/22/business/energy-environment/22SHIP.html

Kalouptsidi, Myrto This is why almost half of cargo ships are sailing around empty, June 16. 2021 Marketwatch https://www.marketwatch.com/story/this-is-why-almost-half-of-cargo-ships-are-sailing-around-empty-11623790696

Say, Mark Rail data marketplace set for beta launch in November, September 29, 2022, UK Authority https://www.ukauthority.com/articles/rail-data-marketplace-platform-set-for-beta-launch-in-november/

Leonard, Ben Nonprofit's app store adds a new twist to health info sharing October 10, 2021, Politico https://www.politico.com/newsletters/future-pulse/2021/10/20/nonprofits-app-store-adds-a-new-twist-to-health-info-sharing-798310

PLAY 23: WHAT DO WE KNOW ABOUT THE PRICE OF GOATS?

BUSINESS DECISION MAKERS WILL FIGHT TO GET THE DATA THEY NEED TO MAKE A DECISION. ONCE THEY HAVE IT, WHAT IS THE BENEFIT OF ASKING FOR MORE AND MORE DATA?

Every few months a farmer sells some goats at the market. She keeps meticulous records of her sales. Looking back over the year, she reviews how many goats were sold each month and how much money she received. That is the data she knows well and has collected.

At the market, each goat is weighed, and its health recorded before it is put up for sale. That data is created and collected by the market authorities, but the farmer never bothered to request it. In aggregate this data exists but the farmer does not realize that it was collected. However, if the farmer were to ask for it, she would understand what drove the weekly price of goats.

There are many other farmers and buyers all trading goats and other animals. The records of the actual price paid for any animal on any day are known by the market. The farmer knows of this data, but she does not have it. When the farmer bought that data from other farmers, she learned how other farmers were making more money by keeping both chickens and goats.

One day all the farmer's goats were bought by a single buyer. The farmer followed the buyer and the goats he'd bought. The buyer walked the goats back to a dairy. At the side of the dairy, a shop sold goat's milk and cheese. The cheese and the milk were shipped to stores in large cities where it commanded a very high price. The farmer was surprised as this was data she didn't know existed and she did not have before, however, with this knowledge she was able to start producing her own cheese.

It's very easy to stop at the data you already collect and analyze, but the big value comes when you keep seeking new sources of data.

Applying the Data Mindset:

Look beyond the data you know and have. Now look deeply into your business processes, what data is being created but not captured. Look outside your own organization to see other data that is collected. Are there uses for that data to create value? How if you had to, would you acquire that data?

The last part is difficult, looking right outside your organization what data might you not be aware of, and you don't have? How might you use that data uniquely?

References:

When Is the Best Time To Sell Meat Goats?
https://familyfarmlivestock.com/when-is-the-best-time-to-sell-meat-goats/

PLAY 24: WHERE CAN YOU SELL YOUR HOLIDAY PHOTOS?

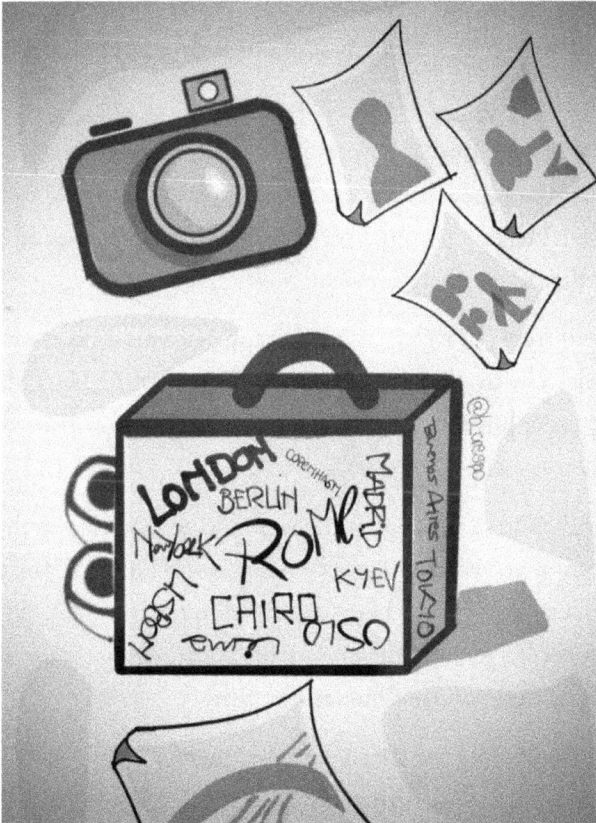

ONLY PROFESSIONAL PHOTOGRAPHERS AND SERIOUS HOBBYISTS ARE PAID FOR THEIR HOLIDAY PHOTOS. WITH A DATA MINDSET YOUR HOLIDAY PHOTOS CAN BE TURNED INTO REVENUE.

A clever technology startup turned social media vacation photos into a mobile application advertising luxury hotels. Designed for world travelers to search for exotic holiday destinations on a popular social media site. They would be shown genuine photos of the area taken by actual visitors, and mingled in, they'd also be shown photos of hotels in the area.

Starting from scratch the team didn't have the budget for professional destination photography. But they needed the content and traffic attract the advertisers. This is a story of how data entrepreneurship and technical genius we combined to find the best photos while respecting personal data boundaries.

Popular social media have sophisticated permissions. The person posting photos can specify who sees those photos: nobody; specific friends; or everybody. The app used the permissions set by each user to show their photos to other users where permission was given.

When intrepid travelers used the app, it would access their photo albums to share with other travelers that used the app. Thus, the app was able to acquire a huge catalog of real traveler images with permission for free.

The attraction of the app for travelers and advertisers was the beautiful images. Computer vision was used to process millions of posted images, algorithms that recognized high quality photography and filtering out photos of people. It used the photo file location metadata and tagged descriptions to find points of interest for visitors.

If you don't have the data you need, there are cost effective ways of either capturing or synthesizing it at scale. Data entrepreneurship at its finest.

Applying the Data Mindset:

Think of a new business model where your organization can generate value to existing or new customers. Is there critical data you don't have to enable the business model? Brainstorm ways of obtaining that data for little or no cost. What would it take for that supply of data to scale and be sustainable for the long term?

References:

What does Jetpac Measure?
https://petewarden.com/2013/12/04/what-does-jetpac-measure/

PLAY 25: WHAT'S YOUR BIGGEST VISION FOR DATA?

JUST ONE PERSON CYCLING TO AND FROM WORK LEAVES A VALUABLE DATA TRAIL. WITH THE RIGHT VISION, THIS CAN LEAD TO PERSONAL DATA STORES AND OPEN BANKING.

As an experiment, a London art school student stalked a bicycle commuter for weeks, logging the route they took between work and home. The student then left notes on the bike advising of faster and safer commuting routes.

The student went on to digitally stalk himself. He approached every company that he did business with and made a formal request to release his personal data. Under the UK's data protection act, a business is obliged to release data for a £10 administration fee. After a long and inconvenient process, he obtained a stack of printed pages of his own personal data. The data was successfully sold online for £150.

These experiments were performed long before programmatic advertising. The student pre-empted a market for data worth hundreds of billions of dollars. This experiment was part of a vision far greater and far more lucrative than the personal data market we know today. It was to create a futures exchange for personal data, one where consumer organizations can buy into your data streams to build high value, sustainable relationships with their customers.

What if your bank paid you in advance for exclusive access to your financial data for a year? It would begin with the transactional data it already had, and you'd grant permission to add financial data from your other providers, credit card and mortgage companies. This is now possible with open banking. The bank would accurately predict exactly your need for financial products. How much should the bank pay you for that data? How valuable would that exclusive access be to the bank?

Applying the Data Mindset:

When you approach a data analysis project, extrapolate it into the distant future. Is there a bigger market for that data or the immediate insight derived from it? What might the data be worth to the right bidder? How can you turn it from a one-off sale to a sustainable stream of revenues?

What model would protect the original owner of the data and provide them some benefit?

References:

Downs, Chris What happens when you sell your personal data on eBay: a first-hand account from a data enthusiast November 17, 2017, Startup Garage https://medium.com/startup-garage-at-station-f/what-happens-when-you-sell-your-personal-data-on-ebay-a-first-hand-account-from-a-data-enthusiast-b8a07fd44155

PLAY 26: HOW SMART IS YOUR HOME?

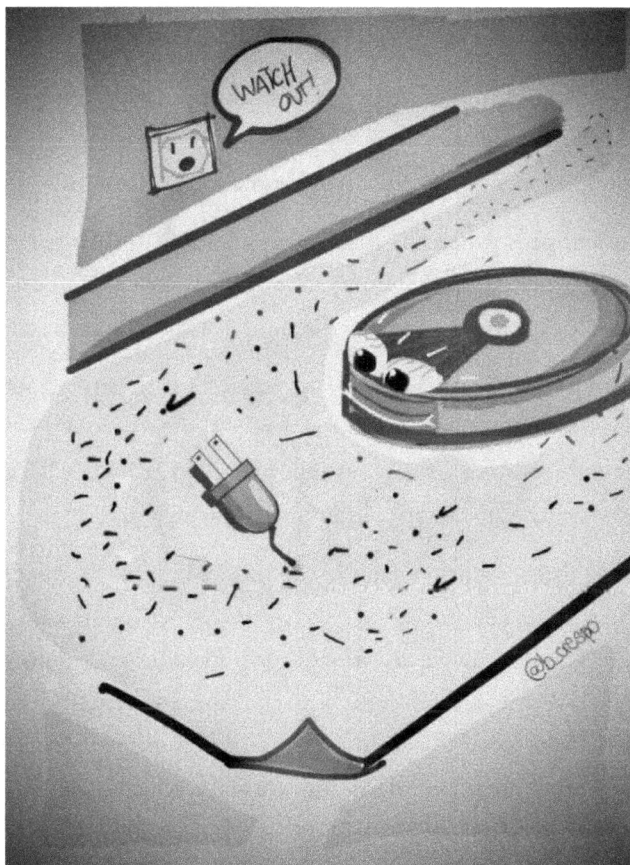

IT'S NOT LONG BEFORE OUR HOMES WILL RESEMBLE THE SCIENCE FICTION OF THE PAST DECADES. JUST WHAT DATA WILL THOSE SMART HOMES BE COLLECTING ABOUT OUR LIVES AND WILL THE BENEFITS OUTWEIGH THE RISKS?

Smart devices around the home like cameras, doorbells and speakers are obviously collecting data. Yet other devices we own like smart thermostats, robotic vacuum cleaners, and connected refrigerators are doing the same. If we were to ask them what they are learning, what might they tell us?

A smart thermostat will tell us the temperature that is preferred and maybe what the weather is outside. Smart speakers can tell what languages are being spoken and when there is a stranger in the house. A robotic vacuum cleaner is creating a dynamic map of the house. The fridge from its door opening frequency can tell when people are not at home or asleep.

What happens when all that data is joined up? Fortunately, new privacy regulations are being introduced to prevent our most intimate movements from being recorded and retained without explicit consent.

However, if the data was aggregated with the homeowner's permission, how can we help the homeowner? Can an intruder be detected, identified, and reported? Can we save the homeowner money for heating or cooling the house? Can we offer interior design advice?

Perhaps there is an opportunity to create a hub for all the data in the house to be aggregated. Smart homes of the future will come with their own data clouds. When an owner sells their home, in addition to selling the land and construction, the sale will come with the data collected from that house. This data will allow the new owners to see equipment specifications, installation and maintenance dates and warranty information.

Applying the Data Mindset:

As you think about your business, are there processes and devices that are just collecting data as they follow the task flow? Here are a couple of examples:

Analyzing the internal email traffic by building a network map will show who the hubs of information are, who is influential, who contributes most.

Analyzing free form written communication from customers, plotting topic and sentiment in correlation with product launches or major events might reveal ways of predicting customer response to the next event.

References:

Briefly describe the advantages and disadvantages of a smart home May 9, 2022, Video Strong News https://www.videostrong.com/news-show-35

Surveillance capitalism: big tech's invasion of our private spaces through smart home technology – part two https://www.cudos.org/blog/%F0%9F%93%B7-surveillance-capitalism-big-techs-invasion-of-our-private-spaces-through-smart-home-technology-part-two-%F0%9F%8F%A0/

PLAY 27: SORTING FREEWAY TRAFFIC

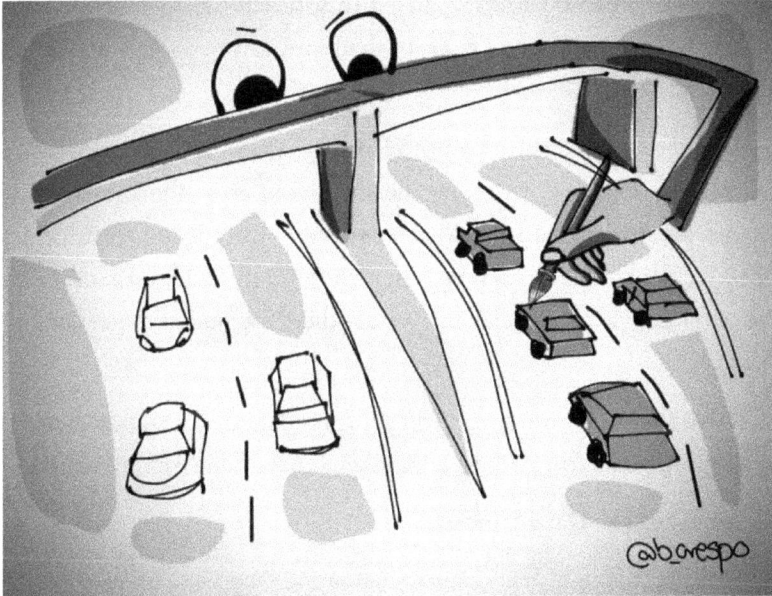

To allow our limited human brains to differentiate the vast number of things we encounter, we naturally group and classify them. To someone with a data mindset, simply sorting anything creates value - even sorting cars on the freeway.

There's a video of traffic on a freeway in San Diego over the course of an hour. The footage consists of cars under a bridge driving in both directions. The curious thing is that all the cars are sorted by color and type. First white sedans, then silver ones, then black and so it goes on. Then it goes from sedans to vans, then buses and finally motorcycles.

The video had been manipulated to sort vehicles by color and type. It's a simple, but fascinating concept that appeals to people who like their data sorted.

The technology to recognize elements within a video feed is now commercially available. Consider the sheer volume of footage now being collected by security cameras. When camera footage is joined up and the video data can be analyzed, there is tremendous information there. There is huge potential value but with that value, there are potential privacy rights infringements.

What increase in value can you get from data that is simply sorted? As you add more data from other sources, what new value can be created?

Applying the Data Mindset:

After receiving a new dataset, first sum, then sort, on any attribute possible. Join the set with new data, align the consolidated data sets, and try sorting it again. The simple act of aggregating and sorting can lead to creative inquiries and consequently, insights. What data in your organization can you try this technique on?

References:

Vimeo: Midday Traffic Time Collapsed and Reorganized by Color: San Diego Study #3 http://vimeo.com/82038912

Kwet, Michael The Intercept: The Rise of Smart Camera Networks, And Why We Should Ban Them January 27, 2021, The Intercept https://theintercept.com/2020/01/27/surveillance-cctv-smart-camera-networks/

Briefcam Video Synopsis Technology: https://www.briefcam.com/technology/video-synopsis/

PLAY 28: WAS IT REALLY BEGINNER'S LUCK?

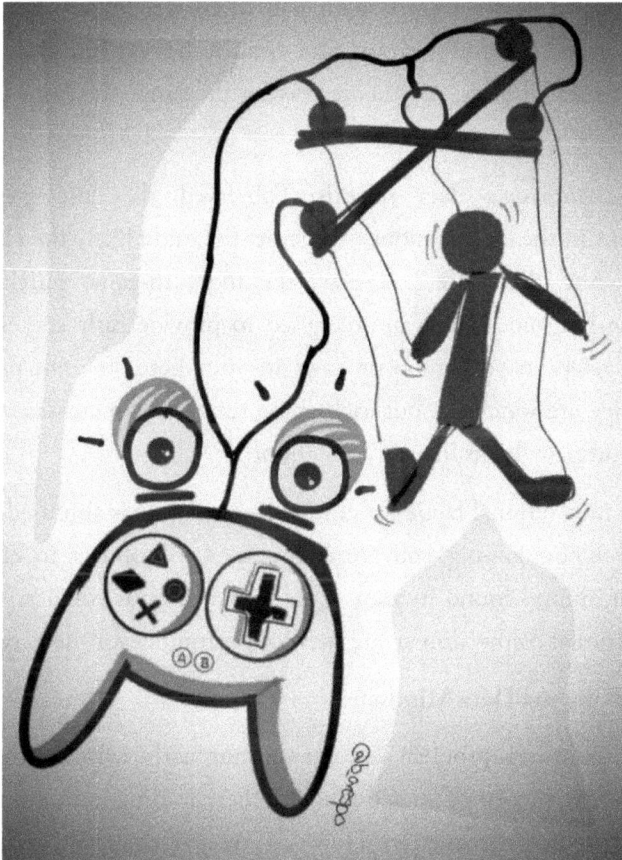

WHEREAS OLDER VIDEO GAMES WERE SOMETHING THAT YOU
PLAYED, TODAY'S CONNECTED MULTIPLAYER GAMES ARE ACTUALLY
PLAYING YOU. WHAT CAN WE LEARN FROM HOW GAMING
COMPANIES THINK ABOUT DATA?

I went to visit a famous online gaming company in San Francisco. It didn't look anything like a traditional software company, more like the trading floor of a bank. This, it turned out was a critical part of its success.

They have a popular gangster game where players fight and rob other players. The first time you play the game your hit rate is uncannily high. The same good luck would be with you when you played their farming simulator game, bountiful crops, and healthy beasts. But this was only during your initial plays. As you are playing the game, the game is playing you.

The company's data scientists observed that the greater the engagement in the early sessions of a game, the more likely the player is to continue. And the more you play, the more in-game purchases and advertising get sold. Games are designed to provide early successes, how many gangsters in your mob or cows in your herd. It's not only those metrics they are watching, but each game runs its own currency which is why the company looks like a trading floor.

This motivational tactic for customers to engage is also used by social networks. Upon joining you are encouraged to connect to colleagues, friends, or alumni found in your contacts list or your contacts' contacts. Each additional connection creates new opportunities for the network.

Applying the Data Mindset:

In the business problem you are working with, what are the metrics that define the success of an activity? Are there leading indicators that can be used to predict those metrics? How can you deliberately drive the leading indicators, even with an artificial intervention like making a game easier for the first few plays?

References

Pascal-Emmanuel Gobry How Zynga Makes Money September 28, 2022, Business Insider https://www.businessinsider.com/zynga-revenue-analysis-2011-9

PLAY 29: THE JOYS OF TIMESTAMPS

WHEN THERE IS A MULTI-STEPPED PROCESS, THERE WILL BE VARIABILITY OF TIME TAKEN OR FLOW THROUGH THE PROCESS STEPS. WHAT CAN THIS TELL US ABOUT EFFICIENCY?

A rideshare driver recorded the time-to-pickup, the bonus payment, and fare she received, for each time of day and date. As 'just a driver', she didn't have access to the APIs to pull the data, and the rideshare company protects the dynamic pricing algorithm. Yet with pen and paper she was able to figure out the times of day for her to get the greatest return for working. With simple analysis this driver created huge insights.

Like our driver, today's software applications keep a log of what happens for diagnostic and troubleshooting purposes. Each time the system does something the activity, the date and time it happens, and the unique identity of that transaction is recorded. These logs were only examined when there was a problem, but the information they contain is golden.

Aggregating a year's transaction logs from every system used by a business process creates a lot of timestamp data. If that data is now sorted by the unique case ID, then we see how quickly processes are completed and the steps that are taken for each process. We would see trends of processes getting stuck at a step, or processes that ended in a failed outcome. This is the science of business process mining, looking at the actual detailed steps involved to find patterns, bottlenecks, and redundancies.

Organizational business processes regularly get stuck or are known to be just plain inefficient. Conversations with customer facing team members show they are acutely aware of the problems. However, senior managers are shielded from the actual day to day pains because they look at reporting summaries. Visualizing timestamp data by process exposes the actual delays, failures, and bottlenecks.

Do you think process owners and decision makers need this insightful view of their business?

Applying the Data Mindset:

Pick a business process that everybody knows to be troublesome. Through speaking to stakeholders, document what the known issues are. Now analyze the process logs to see which transactions are taking longest or getting stuck. What do problematic cases have in common? Are there processes that vary wildly from the norm? How can the inefficiencies be remedied?

PLAY 30: THE VISUAL CUE
OF A BUCKET OF WATER

THE SIMPLEST VISUALIZATIONS CAN IMPROVE COMPREHENSION OF
DATA. TODAY'S TECHNOLOGIES ALLOW ANALYSTS TO QUICKLY
CREATE ELABORATE GRAPHS AND CHARTS, BUT SOMETIMES THE
MOST USEFUL INSIGHTS CAN BE AMAZINGLY SIMPLE.

As a water saving tip, leave a bucket under the running shower while you're waiting for the water to warm up. The result is twofold, you get a bucket of clean water and, also a good understanding of how much water you've wasted.

With the word processor spellcheck feature turned off, a document with spelling errors looks identical to a perfect document. Word processors visually highlight misspellings and grammatical errors to quickly draw attention to them.

It's worth remembering two important design principles, first asking what the system needs to accomplish, then making it easy for the user to take action. For example, positioning of door handles can imperceptibly tell you whether to push or pull. When designing for data you can apply the same principles.

A mobile navigation application aggregates all the location and speed data from cars using the app. These are converted into a real-time traffic map that is overlaid onto the drivers' navigation screens. If the system detects a delay on the intended route, the app presents one simple thing: a new route option. The driver does not get exposed to all the background data and complex calculations. They simply see the instruction to make a change and the updated journey time. Presented at decision time, using relevant and timely data, the app gives the user a simple decision to make.

Applying the Data Mindset:

As you are analyzing data and looking for an insight, step back and consider the user's story. What insight and action would make the user successful? What critical information does the user need to make the simplest decision? How can you present that information, and the ability to act on it, just at the right time?

References:

Morgan, Jesse Russell Intro to UX - The Norman Door December 28, 2018, UX Collective https://uxdesign.cc/intro-to-ux-the-norman-door-61f8120b6086

Godin, Seth This is broken TED https://www.ted.com/talks/seth_godin_this_is_broken?

PLAY 31: DIGITAL IDENTITY IN THE METAVERSE

WHAT HAPPENS TO YOUR SOCIAL ACCOUNTS WHEN YOU DIE? THIS IS AN OPENING QUESTION IN A CONVERSATION ABOUT DIGITAL IDENTITY, AS WE TRY TO RESOLVE BETWEEN DIGITAL AND PHYSICAL PEOPLE IN AN ONLINE-OFFLINE WORLD.

After a person dies, we continue seeing their social media pages unless they are deliberately taken down. Bank accounts and financial data also remain in various systems until they're archived. The person may be dead, but their data lives on. However, if our data is erased or compromised by identity theft, living in the modern world becomes a challenge.

A person's digital identity is now critical. Without it we cannot travel, get a job, or buy property. Providing verifiable digital credentials has replaced carrying a physical ID card or passport. We now rely on our mobile device to either authenticate us by PIN, fingerprint, or facial recognition, and that is used to digitally sign a request for permission. But even this is due to change.

In the physical world, we had one body and one identity. In a digital world, one person can have many digital identities, often for a single service. For target marketing, trying to identify a single person across multiple identities requires work. Maintaining separate business and personal online identities reflects a person's different behaviors. In these cases, should we consider a market segment of one or even a fractional person?

The rise of purely digital metaverses allow avatars to act independently of, even anonymously from their owners. An avatar can be a person's alternate persona or even the persona owned by multiple real-world identities. Is an avatar the equivalent of a corporation, as a brand that unites multiple entities?

When analyzing people's digital personas, we should consider the new relationship between physical and digital people. Can a digital persona spend the money owned by a physical person? If we collect data on a digital person acting online, is there sense in joining it to the data of a physical person living in the real world? All questions that need answers.

Applying the Data Mindset:

Does your business deal with individuals as a part of its product, as customers or as associates? What insights are you looking for about the individuals? Where is your data coming from and what are the rules that determine a physical person versus their online personas.

How do you analyze their behaviors or transactions and how do you want to influence their actions? How best can you carry out the analysis – as a group with independent behaviors, as a single person, as a facet of an individual or as a group of people acting through a single avatar?

References:

Parker, Pamela. "What is identity resolution and how are platforms adapting to privacy changes?", *Martech* June 1, 2022
https://martech.org/what-is-identity-resolution-and-how-are-platforms-adapting-to-privacy-changes/

Wolfson, Rachel. "Reinventing yourself in the Metaverse through digital identity, *Cointelegraph,* August 11, 2022
https://cointelegraph.com/news/reinventing-yourself-in-the-metaverse-through-digital-identity

PLAY 32: AN INVESTOR PITCH FOR EMAIL

AS ANNOYING AS WE FIND OUR INBOXES, WE CAN'T SEEM TO SHAKE THE EMAIL HABIT WHATEVER NEW APPLICATIONS WE TRY. WHAT IS IT ABOUT EMAIL THAT MAKES IT AN INDISPENSABLE PART OF OUR COMMUNICATIONS FABRIC?

The question "Is there anything to replace email?", posted on an online forum prompted a flood of responses. Instant messaging, collaborative workflow with message threads, document signing applications, and many more ideas were discussed. Although each idea solved specific issues, none covered everything email is used for.

What would an investor pitch for email look like today? The target market is everybody socially and in business. Features designed for every device, every vertical industry, and every business function. The investor conversation would end very quickly with some advice to tone down the ambition.

Yet despite decades of alternative technologies being launched, email continues to prove its utility. But this is because email is a highly effective protocol rather than an application. This gives it wider applicability, which also makes it a poor venture investment.

Many data infrastructure products do not produce a return but enable other processes that create internal and external value. Investing in data quality, master data management or semantics and indexing all make the data easier to use and lower the costs to develop other applications.

Applying the Data Mindset:

Is there a project or new application that would be improved with access to better quality data? Are there small improvements in data collection or structuring of data that can be implemented as part of that project? How could the improved data be used by other applications inside your business, or with suppliers or clients?

References:

LaFrance, Adrienne. "The Triumph of Email, Why does one of the world's most reviled technologies keep winning?" The Atlantic January 6, 2016, https://www.theatlantic.com/technology/archive/2016/01/what-comes-after-email/422625/

PLAY 33: DATA AT THE GAS STATION

EVERYWHERE WE LOOK THERE IS DATA AND THAT CAN BE TURNED INTO INSIGHTS. THE NUMBER AND COLOR OF LEAVES ON A TREE WILL TELL YOU WHAT SEASON IT IS. IF YOU CAN SEE THE DATA, THEN YOU CAN FIND WAYS TO CREATE VALUABLE INSIGHTS THAT TURN INTO DATA PRODUCTS.

The next time you see a gas station take a few minutes to see the data. If you were an analyst working for the gas station company, what data would you collect and what could you use it for?

The most obvious are fuel sales and in-store sales by day or time of day. What grade of fuel, what products are being purchased. How many cars go through the car wash or stop for air or water? What is the price of everything and how does that affect demand?

Look at the cars that pass through, what are the makes and models, how old are they? Do certain makes come more often? How many people are in each car? Consider additional points, such as the license plates of customer cars passing through and the make and model of each customer vehicle.

Are there children in the car? Does that effect how long the car stops for or how much is spent in the store? What direction do the cars come from and where do they go afterwards?

What about the weather, how does that affect traffic and sales? Do fuel prices affect the types of cars that come in or times of day? Are there other gas stations close by, how do their fuel prices affect sales?

Can you count the number of cars passing by the station without entering? How many times does each car pass the station for each time it comes in for fuel? What can be done to influence those numbers?

Combining data from different sources allows us to answer more questions. Seeking creative ways of capturing data will lead to new ways of solving problems.

Applying the Data Mindset:

Leave no stone unturned when thinking about where you can collect data. Think about data in the following categories: (1) Data you already have; (2) Data you know of, but don't have (3) Data you didn't know you had; and (4) Data you'd never thought about before and do not currently collect.

Draw a 2x2 grid and list the data sources by category in each quadrant. Once the list is complete, brainstorm the insights available if all data sources were accessible. Then find ways to get the missing data that could deliver benefit when analyzed.

PLAY 34: JUST HOW BIG IS AFRICA?

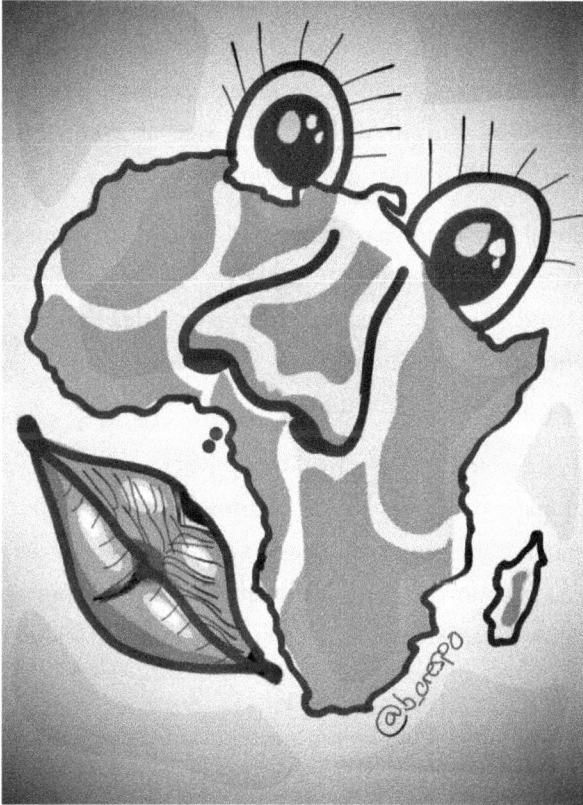

WHEN WE LOOK AT A WORLD MAP, WE PERCEIVE SOME COUNTRIES OR CONTINENTS TO BE SMALLER THAN THEY ARE. PROJECTIONS IN MAPMAKING HAVE A TREMENDOUS IMPACT ON PERCEPTIONS AND THEREFORE POLICY. HOW CAN WE MAKE MORE REPRESENTATIVE VISUALIZATIONS?

Africa is rich in natural resources, boasting 33% of the world's diamonds, 80% of coltan, and 60% of cobalt. It is rich in oil and natural gas, and in manganese, iron, and wood. The arable lands of the Democratic Republic of Congo are capable of feeding the entire continent. Africa has 1.3B people spread over 30M km2, whereas China has 1.4B in 9.6M km2.

Africa is bigger than all of Europe, China and the United States of America combined. Yet, when you look at Africa on a standard map or globe, it looks about the same size as North America. Next to Russia, it looks half as wide, but in reality, it's twice as wide. How has this visualization affected Africa's 1.3B inhabitants?

Traditional map projects have under-projected countries in the Southern Hemisphere. From those maps we have been misinformed that Africa and South America are not as large as Asia, Europe, and North America. These misperceptions have shaped our world view and the importance we give to certain countries.

What if we were to draw a world map where the size of each country was proportionate to its population, a cartogram? What about if we drew the map where the size of the country was proportionate to the production of food in each country? Using different sorts of mapping projections, would we think differently when it came to making decisions about those countries?

"The most powerful weapon in the hands of the oppressor is the mind of the oppressed" ~Steve Biko

Applying the Data Mindset:

What geographies in your organization or market can be visualized as an area map (e.g. sales regions)? Instead of size being proportional to area, what else could you use (e.g., sales growth, total number of months before contracts expire)? How might that change how you deployed resources to those territories?

Instead of geographies, what else could be drawn on an area map (e.g., product categories or products)? What would the area of each product be proportionate to (e.g. product development expenditure)? What happens when you add a relief feature, for example annual sales and are able to visualize the map in 3D, what would you expect to see?

References:

Roser, Max. "The map we need if we want to think about how global living conditions are changing", Our World In Data September 12, 2018 https://ourworldindata.org/world-population-cartogram

Wan, James. "Why Google Maps Gets Africa Wrong" The Guardian April 2, 2014 https://www.theguardian.com/world/2014/apr/02/google-maps-gets-africa-wrong

Raven-Ellison, Daniel "The Netherlands in 100 seconds" YouTube October 23, 2019, https://youtu.be/v0AP18DjLA0

PLAY 35: ASKING A BETTER QUESTION

BUSINESSES SPEND A LOT OF MONEY ACQUIRING NEW CUSTOMERS. IT'S A SHINY METRIC AND NEW CUSTOMERS ARE EXCITING. BUT WHAT ABOUT THE CUSTOMERS THAT JUST KEEP COMING BACK, TIME AFTER TIME. HOW DO YOU INVEST IN THEM?

There's no shortage of data in business, and your competitors will generally have the same data as you do. Therefore, all competing organizations will be looking to find the same efficiencies and routes to winning more customers. Which means they end up asking the same questions.

Questions like: What is the average spend for any customer? What products are they buying? How much does it cost to acquire a new customer? What does it cost to retain them? How can we grow our loyal customer base? What can we do to encourage a customer that purchases once to come back?

Organizations all asking these questions, analyzing similar data with similar results, will follow similar strategies.

What happens if you ask a fundamentally different question? How many customers, that we spent money to acquire, only come back once? Twice? Only three times? Who are our very most loyal customers? How many times do they repeat purchase? How many friends does each of them recommend to? What can we do to encourage and retain our most valuable customer? How can we change our product or service to increase the value of those customers? What if we were able to put those most valuable customers onto our balance sheet? How do we invest in our customer base?

These are powerful questions that help us focus on what needs to be done to take care of the customers to whom we matter.

Applying the Data Mindset:

Look at all the standard reports in your organization, especially the executive KPIs. Are your competitors looking at the same metrics? Are there other ways of looking at your business that your competitors might not be considering?

This might require a deeply introspective look at your business, understanding customers, staff, utilization of space or equipment, the sales cycle, the customer lifecycle, the employee lifecycle.

Of each focus area, ask as many new and diverse questions as you can, write them down and then see which ones might give you some advantage over your competition.

References:

Hevizi, Tamas "Rethinking Customer Value Creation with Peter Fader" Digital Value Creation July 30, 2020 https://www.digitalvaluecreation.io/interviews/rethinking-customer-value-creation-with-peter-fader

Val Rastorguev, Peter Fader and Daniel McCarthy 'How to undervalue your business by 50% in one easy step' ThetaCLV October 4, 2019, https://thetaclv.com/resource/how-to-undervalue-your-business-by-50-in-one-easy-step/

PLAY 36: GRANULES OF ACCOUNTS PAYABLE DATA

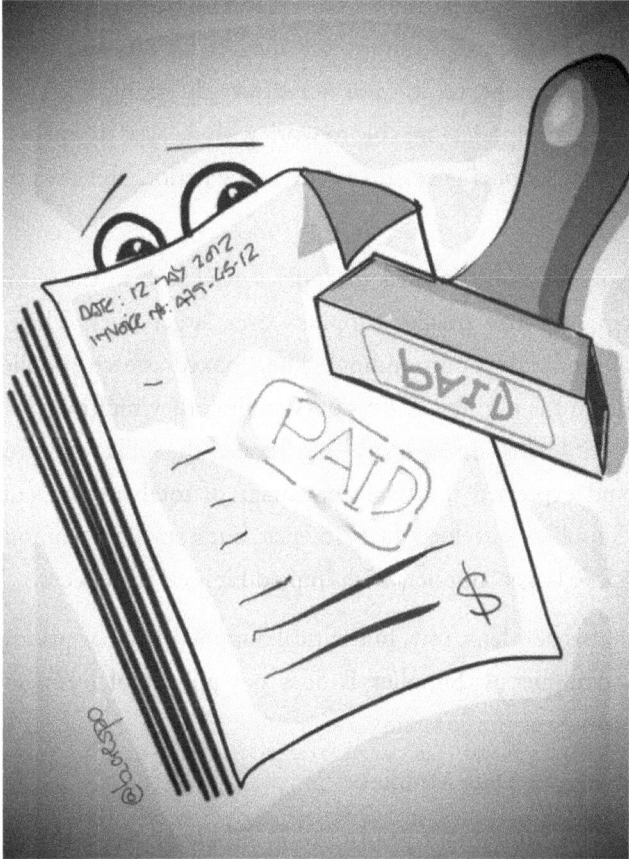

WHEN REPORTING ON A LARGE VOLUME OF DATA, WE ARE QUICK
TO AGGREGATE AND SUMMARIZE. WE CREATE AVERAGES THAT
TELL US HOW WE ARE DOING. BUT VULNERABLE INDIVIDUALS WITH
SPECIAL NEEDS CAN GET LOST IN A WORLD OF AVERAGES.

A multi-national organization depends on a vast network of suppliers. Each year it pays billions of dollars in invoices for their products and services. Suppliers ranged from stable multi-nationals to small local suppliers whose survival depended on on-time payments from their clients.

As per supplier contract terms, the company tries to pay invoices on time. To minimize late payments the organization uses 'days outstanding' to prioritize. They created reports on invoices by age, and the oldest are expedited for payment, yet this was not effectively taking care of the smaller suppliers. Bigger suppliers are able to work with delayed payments, however when a smaller supplier goes unpaid for a long period, this puts their entire business at risk.

The organization decided to apply a new lens to the problem that would help prioritize smaller suppliers over larger ones. They used the concept of 'dollar days outstanding'. They looked at each supplier invoice and calculated the number of days outstanding, they multiplied this by the amount owed giving 'dollar days' They totaled the dollar days from all the invoices and expressed this as a percentage of total invoice value to the supplier. Suppliers with lots of late invoices but a small total amount would be paid before larger suppliers with unpaid high value invoices.

With this new lens, tiny, but critical Supplier B can be prioritized over the larger Supplier A. Supplier B does not go out of business and can continue to service the company.

Applying the Data Mindset:

When encountering a prioritization problem, access the data at the lowest level of granularity. Then apply a new lens to the data to discover a new methodology to assign priorities.

PLAY 37: ROCK, PAPER, SCISSORS FOR DIGITAL BUSINESS

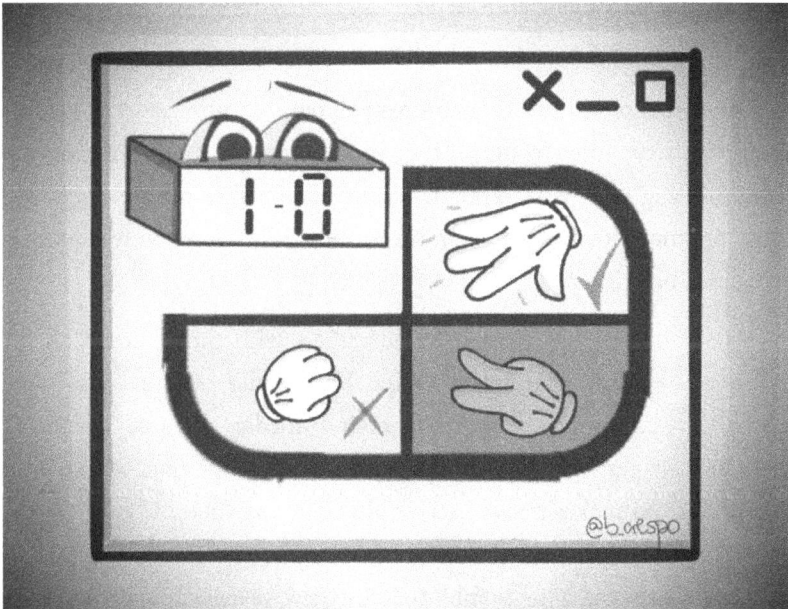

THE GAME OF ROCK, PAPER, SCISSORS PROVIDES A USEFUL
METAPHOR FOR MOVING FROM AN ANALOG TO A DIGITAL AGE.
HOW CAN WE USE THAT TO ENSURE THAT NEW PROPOSITIONS ARE
DIGITALLY NATIVE RATHER THAN ANCHORED IN THE PAST?

Until recently paper documents such as ledgers or passports recorded important information. Scissors and other machinery manipulated physical objects. The persistent rocks of our lives were the material investments with physical assets – houses, cars, or land.

If those were our pre-digital age "rock, paper, scissors", what are they now in our connected world of data? Information has migrated from paper to the cloud. Critical tools include computer, mobile and digital technologies. And ownership of hard assets has been replaced with as-a-service models, leaving relationships as the persistent asset.

An online photo printing company allowed customers to upload and store unlimited photos for free if they spent a certain amount of money on photo printing each year. They failed to predict that people don't print photos anymore, they share them on social media, along with videos that just cannot be printed.

The company was trusted by customers with their photos, so how would they leverage that relationship? They could offer a secure cloud backup for customers' important digital media, photos, certificates, documents, paperwork. They could create technology that would index, organize it, and make it easily accessible from anywhere. Creating a slick user experience would be an improvement over the consumer's inbox as a storage mechanism. This would create a new revenue opportunity that would scale with digital adoption.

When the paper and scissors change, then the rocks provide a new foundation for the proposition. Based on the trusted relationship with the company people would rely on them for storing and distributing important digital artifacts.

Applying the Data Mindset:

Carry out a SWOT analysis on your existing business. What are the prevailing trends in the social and business environment?

What are the transactions (paper)? What are the tools (scissors)? What are the persistent assets (rocks)? How are these being transformed in a new digital business landscape?

How can your business ride the waves of change by redefining the transactions, the tools, and the assets.

References:

Dias, Gam 'Rock, Paper, Scissors for Digital Business' Imagine a World, January 1, 2013, https://www.realtea.net/rock_paper_scissors

Dias, Gam Answer to the question: Who was the original team behind Snapfish, the photosharing and photoprinting site, 2015, Quora https://www.quora.com/Who-was-the-original-team-behind-snapfish-the-photosharing-and-photoprinting-site/answer/Gam-Dias?

PLAY 38: THE 2AM TAXI RIDE HOME

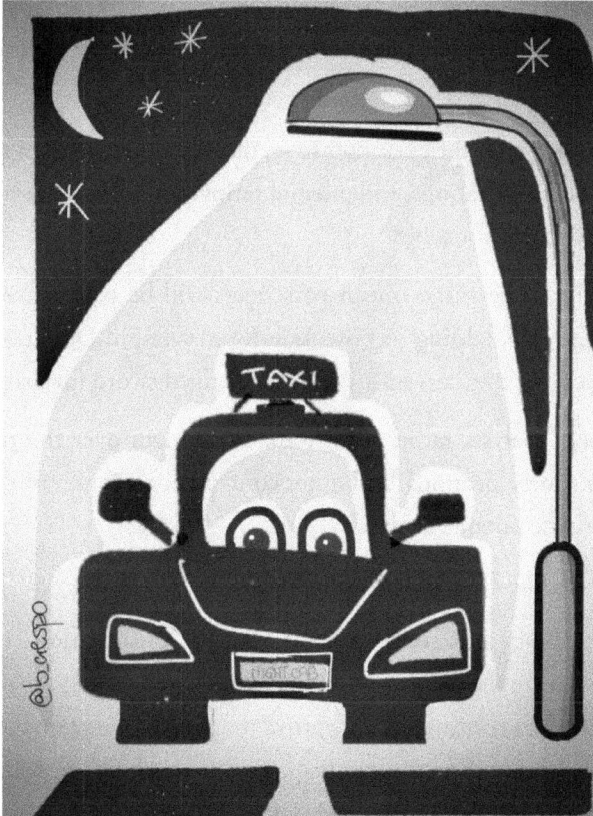

FEEDBACK LOOPS ALLOW ORGANIZATIONS TO CONTINUOUSLY
IMPROVE THEIR PROCESSES AND CUSTOMER SERVICE EXPERIENCE.
GOOD PROCESSES COLLECT FEEDBACK DATA ALMOST
TRANSPARENTLY AT CUSTOMER TOUCHPOINTS IN THE PROCESS
ITSELF

Rideshare companies use driver ratings as a powerful way of changing behaviors. They provide a weekly report back to each driver on how they were rated by their passengers. Dismissing drivers with persistently low scores creates a culture where drivers strive for high ratings.

This data is vital to service quality, so the system ensures that a passenger cannot book a new ride until they have rated their last ride. The rideshare firm built data collection right into the service itself as a design feature.

When analyzing the customer score data by time, the 2am shift shows on average far lower ratings. Judgmental ratings with low scores come from passengers who are not sober.

Late night fare surges mean passengers will be paying more for the same daytime ride, adding to potential dissatisfaction. Despite the higher earning potential, the 2am shift is a double-edged sword for drivers.

So good analysts rather than look at averages over the day, would account for times, locations, and other contextual attributes of each ride to provide a truly equitable analysis.

Applying the Data Mindset:

Think about a business process in your organization, internal or external. What are the feedback loops where you collect data, analyze it, and use the insights to continuously improve the process? Are you collecting all the data you want at the appropriate quality? If not, how can you build data collection right into every customer touchpoint?

When analyzing the performance feedback data, are you considering every context where that data is collected? Break the business scenario down into its component parts.

Ask yourself how these individual components interact and to determine possible variables that may impact performance. Look at time, look at cohorts of users, look at the volume of activity at the time, the weather, and the economy. Invite other stakeholders in the process with different backgrounds or skills to think about why performance might vary.

References:

Cook, James 'Uber's internal charts show how its driver-rating system actually works' Business Insider, February 11, 2015, https://www.businessinsider.com/leaked-charts-show-how-ubers-driver-rating-system-works-2015-2

Schwager, Andre and Meyer, Chris 'Understanding Customer Experience', Harvard Business Review (February 2007), https://hbr.org/2007/02/understanding-customer-experience

PLAY 39: SEEING THE WOOD FOR THE TREES

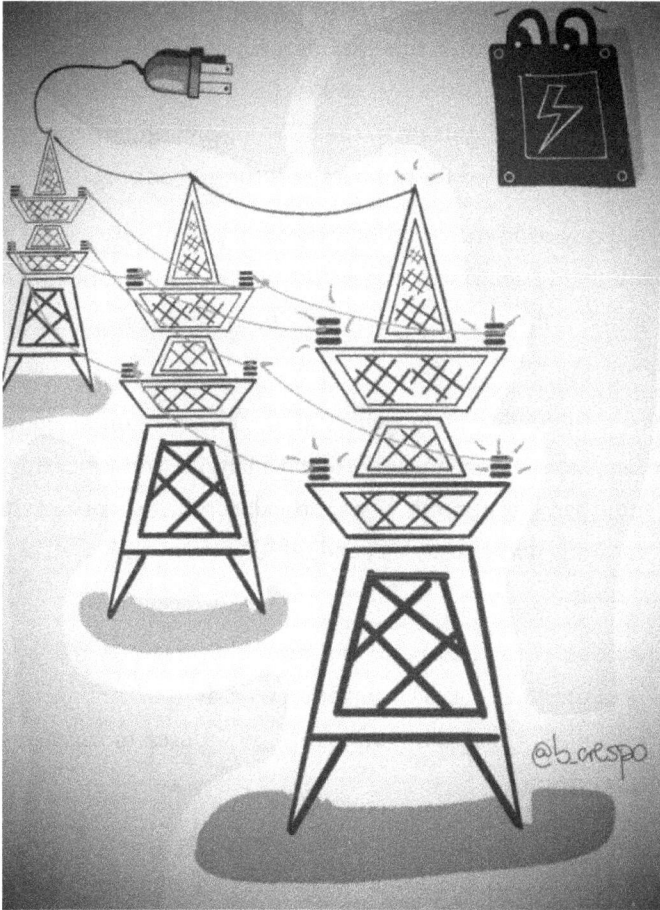

A STORY OF ANALYZING SMART METER DATA HIGHLIGHTS THE IMPORTANCE OF ACTIVELY SEEKING THE INPUT FROM BUSINESS STAKEHOLDERS WHO HAVE EXPERIENCE IN THE PROCESSES.

Smart meters track the usage of utilities and can be read remotely. An experienced data science team was asked to analyze smart electricity meter data. Although experienced in data and analysis, the team was unfamiliar with electric utilities and power distribution management.

The meters recorded power usage in 15-minute increments, as opposed to conventional meters, which provide readings annually or biannually. The increased granularity of analysis led one team member to discover that utility customers could be categorized as follows: (1) customers who turned lights on in the morning and off in the evening; and (2) customers who turned lights on and off throughout the day.

Upon presenting their findings, the data scientist was asked why the data science team had wasted two weeks of the engagement to advise the utility company that the company had both commercial and domestic customers.

Smart meters also know where they are, sometimes sending geo-location data back with the reading. When analyzing consumer smart meter data for suburban single-family homes, the data scientist noticed clusters of meters physically located at the end of the street rather than at each individual house.

Consulting with the operations team they realized this was not a network inaccuracy. Using the installation records, they pinpointed a group of service engineers that did not follow procedures to save time when installing the meters.

Amazing insights can be discovered using granular analysis; however, the sheer depth of this analysis can sometimes obscure the most obvious truths, including truths already well understood by business users, customers, and partners. Go into analysis projects with an open mind and be prepared to listen.

Applying the Data Mindset:

Are you faced with a new analysis project with data that you are unfamiliar with? Begin by reading around the subject area to understand how the metrics are collected and analyzed, look for anecdotes of anomalies. Then develop your hypothesis and share it with business stakeholders that will increase your understanding. Ask stakeholders what additional data they would want to have to increase the value of the analysis.

References:

Akhdar, Wassim 'Where Am I? — Are Your Assets Self-Aware of Their Location on the Grid?' Itron May 10, 2022
https://www.itron.com/na/blog/industry-insights/are-your-assets-self-aware-of-their-location-on-the-grid

Sirolli, Ernesto 'Want to help someone? Shut up and listen!' TED Talks,
https://www.ted.com/talks/ernesto_sirolli_want_to_help_someone_shut_up_and_listen?language=en

PLAY 40: REPLACING COSTLY SURVEYS WITH SOCIAL MEDIA ANALYTICS

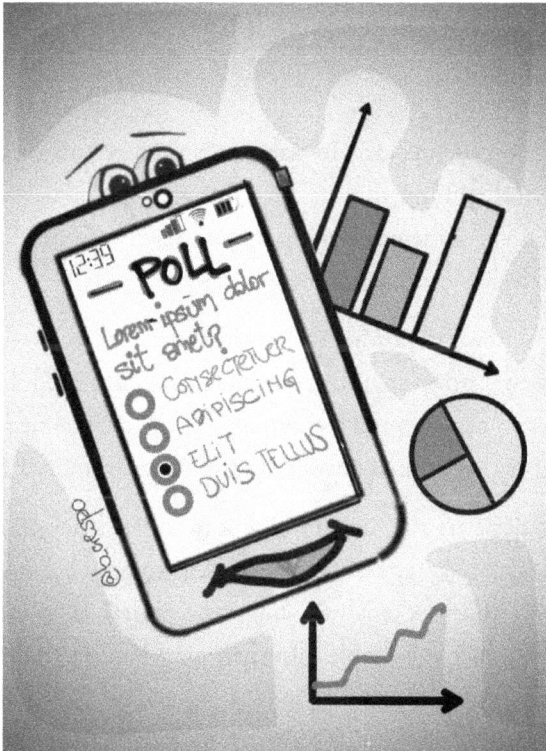

When presented with a business problem, we seek quantifiable and structured data for analysis. As this play from the video games industry shows, analyzing unstructured conversational data can save time and money.

Production of a tier one video game can easily cost hundreds of millions of dollars. To trim this monumental expense a production company was reconsidering the inclusion of a well-known sound technology in its latest game.

The product manager knew that most gamers' systems had high quality audio and assumed that the expensive sound technology was redundant. To support this type of decision they'd contract a market research agency to run focus groups and deploy global surveys to their panel members. This would be a costly and time-consuming project, 6 months and 6 figures!

In this case they decided to use real-time text analytics to monitor social media and gaming forums for customer sentiment. The product manager wrote a genuine post asking the most active forums to voice their opinions on removing the sound technology.

Within 12 hours of posting the product manager received a conclusive result. A huge negative spike in sentiment was observed regarding the potential removal of the sound technology. The game retained high quality sound and the team got their answer almost instantly and for free.

Conversations are the most prolific form of feedback when collecting data from people. This type of textual content might be unstructured and unsuited for spreadsheet analysis, but fast natural language processing (NLP) algorithms allow valuable information to be extracted from the raw data.

Applying the Data Mindset:

In a situation where you are researching a market, the text or unstructured data provides a rich source of information if you understand how to turn qualitative information into quantitative.

Many businesses collect feedback for the purpose of continuous improvement. Social media provides a free source of feedback in the consumer space, which can be leveraged for critical product, service, market, and customer insights. It also presents an opportunity for continuous learnings in the B2B market.

Creating and nurturing feedback loops for your product or service can provide timely and valuable insights, it can even predict nightmare scenarios.

PLAY 41: THINK LIKE A DATA NATIVE

THE GENERATION THAT HAS GROWN UP WITH SOPHISTICATED
APPLICATIONS WITH EMBEDDED COMPUTING HAVE FAR GREATER
EXPECTATIONS OF A USER EXPERIENCE. HOW SHOULD WE EMPLOY
AI TO SATISFY THEM?

We are familiar enough with the term Digital Native, the generation that grew up with computers. Today's "data natives" have even greater expectations. Their "smart" world seamlessly adapts to their taste and habits.

While digital natives were most concerned with what they can do with technology, data natives are more concerned about what that technology can do for them. To do that, intelligent machines act on data.

Digital natives program their thermostat. Data natives expect the thermostat to program itself.

Digital natives use a coffee chain mobile app. Data natives want the app to know their favorite drinks, and when to suggest a new one.

Digital natives use a networked baby monitor. Data natives expect their baby monitor to know if the crying is normal based on millions of other babies' behaviors.

To build these smarter applications requires machine learning that will be able to present options to human users and see what is selected. Then evaluate the outcome to refine what is presented the next time; exactly the same way web search engines get better at making your search results more relevant.

However, in using these augmented intelligence technologies, we need to be responsible in how those systems collect data that is representative of all groups and be careful that we do not program the algorithms with our own biases.

Applying the Data Mindset:

Are you designing a new system or business process? Thinking like a data native, consider how that process could simply be better by understanding, predicting, or recommending things to its users. List as many ways as you can.

Then work out what data you'd need to make those ideas a reality. Is the data readily available, if not can you find it? Do you have enough data, are you allowed to process and share it? How can you create a smart system that does not spook your customers?

References:

Frazier, Mya 'The Data Driven Parent' The Atlantic May 2012 https://www.theatlantic.com/magazine/archive/2012/05/the-data-driven-parent/308935/

Nash, Adam 'My letter to Starbucks mobile', Adam Nash Blog August 27, 2013 http://blog.adamnash.com/2013/08/27/my-letter-to-starbucks-mobile/

'The Responsible Machine Learning Principles', The Institute for Ethical AI & Machine Learning https://ethical.institute/principles.html

Broberg, Brad Unparalleled in parallel reality Microsoft Alumni Network https://www.microsoftalumni.com/s/1769/19/interior.aspx?sid=1769&gid=2&pgid=1577

PLAY 42: WHERE TO FORTIFY PLANES

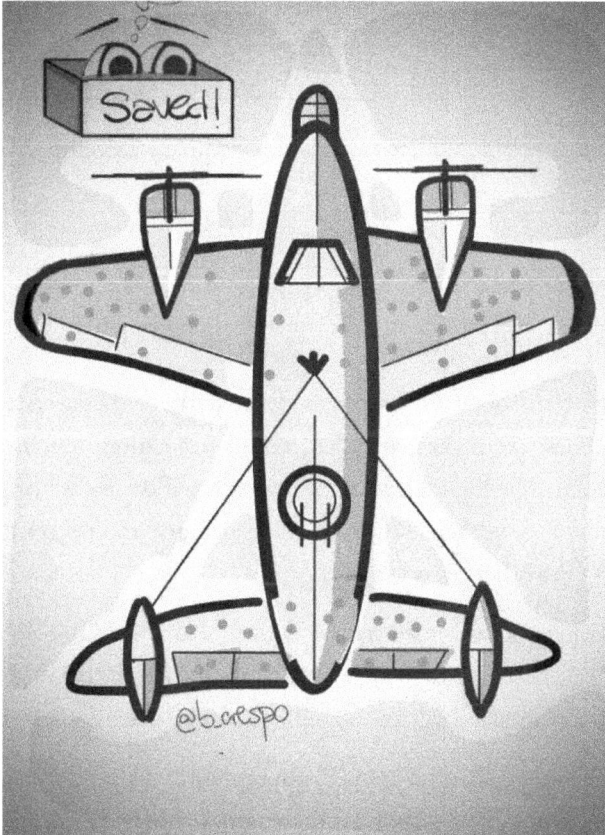

When analyzing data to solve a problem make sure that you're getting all the data, not just a skewed sample of data points that were counted. This play about bullet holes on planes illustrates survivorship bias in data analysis.

War is an expensive undertaking and the military was seeking cost savings. In the heat of World War II, a lot of American planes were coming back riddled with bullet holes. There were distinct patterns to where those bullet holes were most dense, with concentrations in the engine, the fuselage, and the fuel systems.

Section of Plane	Bullet holes per square foot
Engine	1.11
Fuselage	1.73
Fuel system	1.55
Rest of the plane	1.8

The patterns observed highlighted the parts of the aircraft being hit the most. These were the parts that would need more repairs when the planes came in. The logical assumption was that if the parts being hit most were reinforced, fewer repairs would be needed and money could be saved by armoring the other parts less.

The military approached statistician Abraham Wald to help them find out exactly where to fortify the planes. They provided the detailed data from the surviving planes and where the bullet holes were concentrated.

Wald observed that the data did not represent the problem. The reason that the data showed fewer hits on the engines is that the planes that were hit in the engine did not return to be counted. And the fact that a majority of the surviving planes had bullet holes in the fuselage provided strong evidence that the planes could tolerate damage to the fuselage.

Wald offered two explanations for the data:

1. Bullets just happened to hit every other part of the plane more often than it hit the engine

2. The engine is a point of vulnerability

Only the second explanation was viable, so Wald's recommendation to reinforce around the engine was put into effect saving many more planes.

This problem is an example of 'survivorship bias', where when trying to identify performance, you are not counting the ones that didn't survive to be counted. At year's end when analyzing a profile of gym members that joined the previous January to determine the best program to increase fitness, you should also count the ones that dropped out in February and March.

Applying the Data Mindset:

Are you analyzing data to determine what factors can influence high vs low performance? When inspecting the data available, ask the question, "Is this all the data possible or is some missing?" Especially when there were more of the cases to start than at the end. How might the missing values skew the outcome?

When examining a business process, map out the process lifecycle and determine both positive and negative outcomes. Did certain participants in the flow fall out of the funnel before the end. Collect and count their data too.

References:

Ellenberg, Jordan 'An excerpt from How Not To Be Wrong' Abraham Wald and the Missing Bullet Holes Penguin Press July 14. 2016, https://medium.com/@penguinpress/an-excerpt-from-how-not-to-be-wrong-by-jordan-ellenberg-664e708cfc3d

PLAY 43: BABIES PRE-LOADED WITH DATA

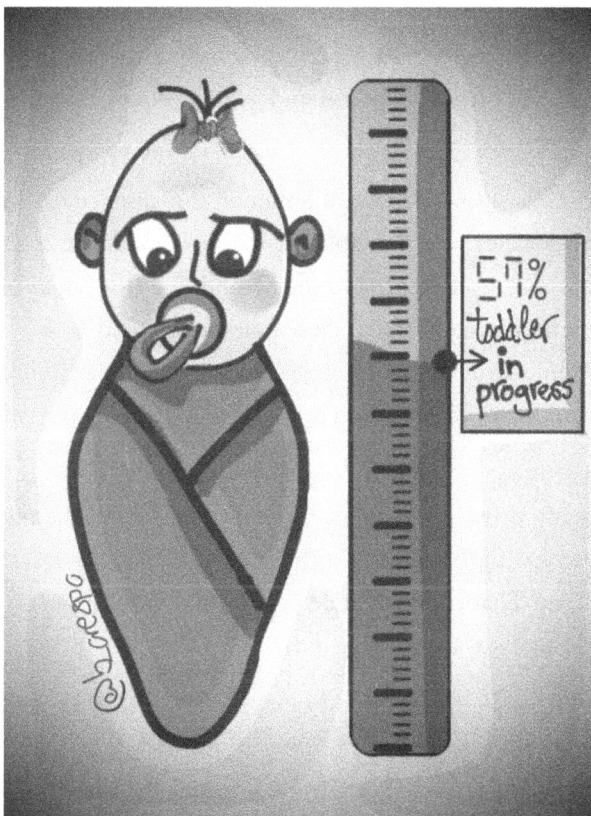

PERSONAL DATA PUBLISHING BEGINS AT BIRTH AND THERE ARE FEW
CONTROLS IN PLACE AS TO HOW IT IS STORED AND SHARED. BY
BUILDING THE NEW PRIVACY REGULATIONS INTO BUSINESS MODELS
WE COULD CREATE A NEW MARKET FOR FULLY CONSENTED
PERSONAL DATA.

Children born after 2021, post-Gen Z, will have their life documented and processed online. Parents and guardians will have provided informed consent without the child's explicit consent. Using only birthday and back to school photos posted in social media, the child's age, home address and biometrics become a commodity in the public data economy.

During pregnancy, parents create online accounts for their unborn children to qualify for free sample boxes of baby products. Once the child is born, they'll add birthdate, name, and gender. Before birth, the data trail has already been laid.

As the child grows, and again under the thin cover of checkbox consent, the retailer will be able to capture the child's size, annual expenditure, and preferred clothing brands. Using this information combined with postcode-based demographics, would it be possible to predict the child's future? Can we go further to predict the child's job, higher education or earning potential? This is where data starts to get creepy.

Embracing privacy and agency over personal data, the retailer could do far more, starting with shifting the locus of control for the family's explicit consent. That data would be kept securely in a digital vault and the parents could update or delete the data at any time.

From this vault, parents could safely share with family members wish lists with accurate sizes. Instead of surreptitiously trawling for data, parents would volunteer their data for the right advantages. This idea of personal data plus full agency over it is a key tenet of Zero Party data.

Embracing the idea of zero-party data and building it into your business model will open up new revenue and profit opportunities.

Applying the Data Mindset:

There are now tighter laws around collection, processing, and usage of personal data. Are you capturing and retaining your customers or employees' personal data? How can you go further than compliance by facilitating the individual to maintain their privacy and gain agency over that data.

How can you design services for data subject to gain more value from their data? What new business model can you derive from this position?

References:

Dias, Gam 'Can you build trust through offering privacy?' Privacy for Profit, October 18, 2020 https://privacyforprofit.com/2020/10/18/why-privacy-for-profit/

Dias Gam, 'Refining Privacy' Joined Up Thinking Blog January 18, 2021 https://joinedupthinking.xyz/redefining-privacy/

Hunt, Elle 'Give up Google, don't hit 'accept all': how to fight for your privacy' The Guardian September 28, 2020 https://www.theguardian.com/books/2020/sep/28/carissa-veliz-intrusion-privacy-is-power-data

PLAY 44: WHAT WEBSITES DO AIR TRAVELERS PREFER?

SOMETIMES IN DATA SCIENCE, THERE ARE TRENDS OR ANOMALIES IN THE DATA THAT ARE UNEXPLAINABLE BY SIMPLY LOOKING AT THE DATA. THIS PLAY DISCUSSES ONE SUCH FINDING AND THE NEED TO DO A LITTLE DETECTIVE WORK.

Digital media agencies use data science to analyze the websites regularly browsed by audiences. An agency targeting frequent flyers used mobile browser data to understand interests and intents.

They used mobile phone browsing data taken from airport locations to isolate a sample of air travelers, their target users. They intended to test a hypothesis that travelers browsed certain news, social media and travel booking sites.

The data scientists were surprised by the results from one major airport hub. The aggregated browsing history showed the most popular interest was a particular group of religious scriptures. This was consistent over months and years so not a meme trend. They could find no easy logical explanation available.

They looked deeper into the data and found the traffic from those sites was found to be from a consistent group of addresses. This was evidence that it was not transient travelers' browsing habits.

To understand this, they had to leave the data and speak to people at the airport. They identified the religious group as being baggage handlers who are all from the same country where that religion was popular. The high traffic was from the quiet periods between plane arrivals when they would get time to study their scriptures.

Applying the Data Mindset:

The data will tell you everything, but don't believe what it tells you without applying critical thinking. Unbelievable statistics or insights should be treated as just that. Furthermore, if you hit dead ends with the data, then don't be afraid to go out and talk to people. Once you have come up with a hypothesis and proved it with the data, do everything you can to disprove it.

PLAY 45: A FRICTIONLESS WAY TO LISTEN TO YOUR CUSTOMERS

REQUESTING AND RECEIVING FEEDBACK, BOTH QUANTITATIVE AND QUALITATIVE. ARE VITAL TO ENSURING YOUR PRODUCT OR SERVICE MEETS EXPECTATIONS. HOW CAN WE ENCOURAGE CUSTOMERS TO LET US KNOW WHAT THEY THINK?

There are plenty of online review sites for customers to rate service and leave a detailed product recommendation. But when do people bother to leave a review? Most often when they're either very happy or very upset. But customers have no guarantee their feedback will be read by the business they're reviewing.

Positive customer feedback is a common KPI, but the collection is designed to hit a target rather than initiate actions to improve the experience.

Can we find better ways for businesses to collect genuine feedback, sensitively and at scale. A post purchase email or the checkout operator asking if you found everything okay are particularly good, customers are not going to tell you much. Chatbots can be better, but they still require effort from the customer without a visible reward.

A popular smoothie company tells the story of how it started in business. "We started [our company] back in 1999 with a dream to make it easier for people to do themselves some good. We took our smoothies along to a music festival, where we put up a big sign asking people if they thought we should give up our jobs to make drinks out of crushed fruit instead. Underneath the sign, we put a bin saying 'yes' and another one saying 'no', then asked people to vote with their empties. At the end of the weekend the 'yes' bin was full, so we resigned from our jobs the next day and got cracking."

The smoothie founders gathered feedback right at the point of experience. Their survey methodology included more than customers at the extremes of satisfaction. It also provided a highly visual and visceral set of results that they could act on the very next day.

Qualitative and quantitative feedback is one of the most important data sources for a business. If you're in the start-up world, then it's crucial to finding product market fit.

Building mechanisms to accurately capture customer feedback and feed that into business decisions is a critical data function.

Applying the Data Mindset:

What business processes touch internal or external customers? What are the range of experiences that you know customers are going through? What mechanisms are in place to collect information about their experience? What percentage of your customers leave feedback? What data is being collected and who uses that data to make decisions?

Is there a mechanism by which you can collect that data while the customer is buying or using the product? How can the data collection be built right into the product experience? Is there anything you can do to inform the customer that their feedback has been heard and some action is taking place?

References:

'Little drinks, big dreams' Innocent Drinks,
https://www.innocentdrinks.co.uk/a-bit-about-us

PLAY 46: MAKING DANGEROUS ASSUMPTIONS WITH DATA THAT IS TOO FAMILIAR

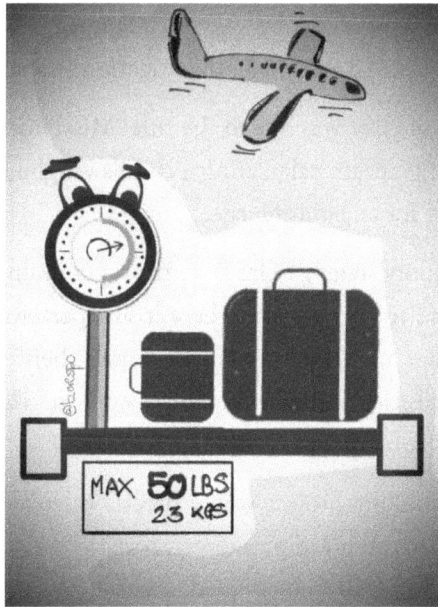

WHERE LARGE SOFTWARE PROJECTS ENGAGE DISTRIBUTED TEAMS, SIMPLE MISTAKES CAN ESCALATE INTO CRITICAL ERRORS. THIS PLAY HIGHLIGHTS WHEN DIFFERENT TEAMS INTERPRET CERTAIN WORDS DIFFERENTLY FOR A CALCULATION.

You may have been asked to move seats before take-off when flying on a smaller aircraft. This is to distribute the load and ensure that the plane is flying within its safety limits. Fatal crashes have been attributed to miscalculated fuel, altitude and thrust due to errors in calculating the load.

Shortly before take-off all commercial airline pilots receive a load sheet, containing information such as the plane's zero fuel weight, weight of fuel and payload. A significant component of the payload will be the passengers, yet this can only be estimated. Also bear in mind that the average weight of passengers is greater today than it was ten years ago.

In 2020, three passenger flights from the UK to Spain were found to be carrying far more weight than documented. One flight of 187 passengers under allocated 38 passengers by over 1200kg. This was classified as a serious incident by the Air Accidents Investigation Branch (AAIB).

The calculation error was due to the title 'Miss' interpreted as being a child passenger. This meant calculating a child's weight allocation of 35kg as opposed to 69kg for an adult female.

The AAIB report stated, "The system programming was not carried out in the UK, and in the country where it was performed the title Miss was used for a child, and Ms. for an adult female, hence the error,". This calculation escaped the software quality assurance checks, fortunately without disastrous consequences.

When planning any project where precise calculations are required, data quality is of paramount importance.

Applying the Data Mindset:

Data quality takes many forms, and it's all too easy to slip up with data that you know well. The more familiar the data, the more you need to take an objective viewpoint and question your assumptions before anything is designed and developed.

When specifying a new system and working with any calculations, particularly for processes that contain high risk, have the team collaboratively create an exhaustive list of assumptions, and then review them to ensure they are valid. Assumptions about data will include expected range values, source, completeness, unit of measure and volatility.

References:

Claburn, Thomas 'Airline software super-bug: Flight loads miscalculated because women using 'Miss' were treated as children' The Register, April 8, 2021
https://www.theregister.com/2021/04/08/tui_software_mistake/

PLAY 47: MORTGAGE APPLICATIONS ANALYZED TWO WAYS

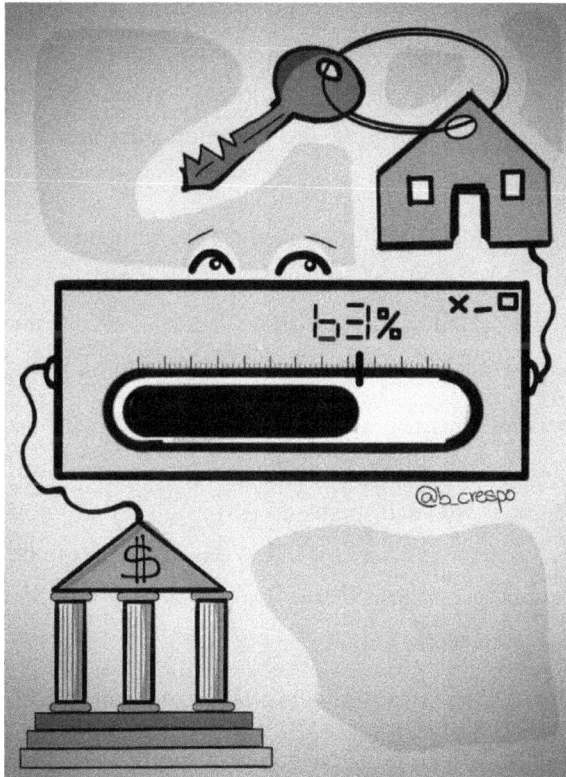

TALENTED ANALYSTS SEE A PROBLEM AND ARE IMMEDIATELY ABLE
TO SEE WHERE THE SOLUTION LIES. HOWEVER, AS THIS PLAY
ABOUT MORTGAGE APPLICATIONS DISCUSSES, THERE IS EVEN MORE
VALUE WHEN INCREASE TEAM DIVERSITY.

A mortgage lender processed thousands of mortgage applications per month each requiring a labor-intensive approval process. Although they approved and funded many mortgages, a high percentage of applications were not funded. Two independent teams were asked to look at the problem.

The web analytics team's hypothesis was that abandoned online applications was the problem. Most applications took a long time to complete and needed multiple iterations. Longer completion times and increased abandonment also happened if the applicant was asked to find additional information. If the applicant got stuck or needed help, they would also quit before finishing and common questions stumped many applicants.

The team came up with valuable recommendations to reduce abandonment: Show the expected time to complete at the start, show a progress bar throughout and even show messages of encouragement; In addition, also provide a list of documents required to fill in the form.

The operations team began by analyzing the actual business process transactions and timestamps, a technique known as business process mining. They saw two main paths taken by submitted applications: one path contained correct applications that received an offer to be funded; the second path contained incorrectly filled applications that needed a call from a service agent to correct.

The incorrect applications were the ones that received a higher percentage of mortgage offers and were funded. Further analysis showed the human touch from a customer service call was a significant factor in mortgage offers being accepted.

The different approaches taken by each team each came up with a different insight and solution, both valuable to the company. This is an example of using diverse thinking to solve problems. If you bring people with different backgrounds and experiences to a problem, you will increase the diversity of thought.

Applying the Data Mindset:

Do you have a difficult business problem to solve? What diverse skills and experiences would increase the number of ways to look at that problem?

Do you have representation from the various business functions, for example finance and marketing? If it's a user interface problem, do you have poorly sighted people represented.

If you can't find representation from all stakeholder groups, then make two attempts at solving the problem. Build a hypothesis and take one approach, then put that aside and force yourself to look at the problem from a completely different angle.

References:

Dobrin, Seth and van der Heever Susara, Putting Diversity To Work In Data Science, Forbes January 24, 2020, https://www.forbes.com/sites/ibm/2020/01/24/putting-diversity-to-work-in-data-science/

PLAY 48: HEDGE FUNDS AND ALTERNATIVE DATA

IF THERE'S LEGAL MONEY TO BE MADE, THEN HEDGE FUNDS WILL BE AFTER A PIECE OF THE ACTION. SO, IT'S NO SURPRISE SOME OF THE INNOVATIVE TECHNIQUES THEY USE TO FIND AND ANALYZE DATA.

The revenue potential in trading stocks can be astronomical. Possessing more timely and accurate information about companies and markets makes all the difference. Banks and brokerages invest in microsecond trading systems to give them competitive advantage. Hedge funds invest in data, here are a few illustrations of the alternative data they track.

A hedge fund bought a freight telematics company to track the movement of goods around the country. This gave it benchmark information as to which organizations were moving goods and therefore the activity levels of manufacturers and retailers in specific industries.

Another hedge fund tracks movements of corporate executives on executive jets using aircraft registries, corporate filings, and flight communications. Unusual patterns, such as a sequence of visits to the head offices of other companies in an industry or even to the home location of a CEO could be the precursor to M&A activity between the two companies.

Hedge funds use satellite imagery of grocery store and shopping mall parking lots. Daily or hourly counts of cars in the parking lots provides them with a proxy for retail sales, particularly in the holiday season.

A European hedge fund purchased the company that maintains the master index of which mobile phone network is responsible for each phone number. Each time a person switches networks, the company registers a customer moving from one provider to another. In aggregate, this shows how each network is growing or shrinking.

Hedge funds have a great interest in electric vehicles (EVs) and how they are powered, particularly to acquire resources such as Lithium and Cobalt. There is a battery plant located in a deserted area, the owners are secretive about production levels.

A hedge fund with an EV portfolio was using satellite imagery to track the trucking activity in and out of the plant to try to predict output, however, a truck could be completely or partially full. To improve the information, they attached sensors to the power lines surrounding the plant. A truck full of charged batteries passing under a power line would generate a magnetic field that could be measured, indicating how many batteries were on the truck.

Applying the Data Mindset:

What market data do you currently research or purchase? Are there alternative ways to find that data or proxies for it? Can you get that information earlier than your competitors? Look up and down the supply chain for dependencies where the data is easily available. Look at published data from annual reports, or customer feedback data for your competitors' products and services.

References:

Bachman, Justin, 'Hedge Funds Are Tracking Private Jets to Find the Next Megadeal' Bloomberg, July 2, 2019
https://www.bloomberg.com/news/articles/2019-07-02/hedge-funds-are-tracking-private-jets-to-find-the-next-megadeal

Kim, Ted. 'Alternative Data for Hedge Funds: The Competitive Edge to Investing, SimilarWeb September 30, 2021
https://www.similarweb.com/corp/blog/investor/asset-research/hedge-funds-use-alternative-data/

'How Satellite Imagery is Helping Hedge Funds Outperform' International Banker, June 26, 2020
https://internationalbanker.com/brokerage/how-satellite-imagery-is-helping-hedge-funds-outperform/

PLAY 49: CHECK YOUR BATTERY LIFE BEFORE CALLING A RIDESHARE

@b_crespo

HAVE YOU EVER BEEN OUT, UNABLE TO CHARGE YOUR PHONE AND WORRIED THAT YOU WON'T BE ABLE TO SUMMON A RIDE HOME FROM YOUR SMARTPHONE? WHAT SORT OF DATA DO YOU THINK IS BEING COLLECTED ABOUT YOU?

When you're ordering a taxi using a rideshare app, try to make sure that your phone is fully charged. The rideshare application is likely to be monitoring your battery life to estimate how desperate you are for that ride. This data is used for surge pricing.

But that's not the only data they collect. Each ride you take adds to the information about where and when you are regularly picked up and dropped off, the messages between you and the drivers and how you rate each driver. Additionally, they collected detailed journey data such as speed and acceleration. This data is used to help the rideshare company improve the service. Having the user's identity and contact details will also protect drivers.

Certain apps are also able to collect sensitive user information, such as sexual orientation, pregnancy, childbirth information, religious, political, and philosophical beliefs and biometric data and genetic information from social media and where your phone number can be used to access your profile. This data is used for marketing and advertising.

It is highly tempting for businesses to collect data from every action object and every movement or transaction. And when combined and analyzed from various angles the value can be game changing for the company or can even shift industries as the rideshare players are doing in the transportation markets.

Most applications that you use on your smartphone are very busy collecting data about the phone, its location and everything else that you do on the phone. Yes, that's creepy and the new privacy laws have been designed to prevent the unauthorized exploitation of unconsented personal data.

The most interesting data is also the most dangerous. Deliberately or inadvertently individuals can be harmed because of decisions taken based on the data collected.

Applying the Data Mindset:

When identifying data that is being used to solve a problem, think carefully about what decisions will be made based on the insight. Think if an individual's or a group's information will be disclosed and how that might negatively impact the individual. The consent notice to collect and process this data should be explicit and require the permission of the user.

If, as in the case of ridesharing (e.g. the user's real time location), the benefits to the users are so great (being able to quickly get a ride) they will consent to their data being used for this purpose. It is up to you to design a service where users willingly contribute their data to enjoy the benefits.

References:

Kingsley-Hughs, Adrian 'Why Uber is watching your smartphone's battery level' ZDNet May 20, 2016 https://www.zdnet.com/article/why-uber-is-watching-your-smartphones-battery-level/

Phillips, Gavin 'Which rideshare apps collect the most data on you' Makeuseof January 28 2022 https://www.makeuseof.com/which-ride-hailing-apps-collect-most-data/

Braw, Elisabeth with Palazzolo, Franco 'How rideshare apps collect and share data: A national security risk' Royal United Services Institute https://static.rusi.org/307_RideShareData-EI.pdf

PLAY 50: IS THERE DATA IN A POTHOLE IN THE ROAD?

ALL PEOPLE HAVE BIASES, MANY OF THEM ARE USEFUL WAYS TO SPEED UP DECISION MAKING. MACHINE LEARNING WILL INHERIT THE BIAS FROM HUMANS THAT USE THE TOOLS, AND SOMETIMES THIS FURTHER DISADVANTAGES CERTAIN GROUPS.

There are all sorts of novel ways to collect actionable data. An older US City on the East Coast had a massive problem with potholes, having to fix around 20,000 of them each year. Even locating them requires coordination and effort.

Smartphone technology allows data to be collected from thousands or even millions of devices, with a minimum of effort from the phone owners. Smartphones come with accelerometers that sense movements and GPS that will locate the phone. When the iPhone was first released, the city devised an app that detected when a phone was in a car and the car went over a pothole and precisely where that pothole was located.

The app was very successful efficiently reporting potholes that they could quickly fix. However, after some time the city noted that the potholes were only getting fixed in the more affluent areas. When the iPhone came out its penetration was focused on higher income individuals. Areas where lower income groups lived were less likely to have their potholes fixed.

When the sample used for analysis is not representative of the entire population, the results can be skewed as we saw with the pothole repairs. This can have a detrimental effect on certain groups that were under-represented.

When designing applications or data experiments, think how that bias can be avoided or mitigated.

Applying the Data Mindset:

Is there a piece of analysis you are planning right now? What data do you need for the analysis and how are you collecting it. What is the business decision that will be made using that data?

Which individuals will benefit from that decision? Are there individuals or groups that might miss out on the benefit, and what would be their loss?

Does the analysis data include enough data from the groups that potentially don't benefit? How can you ensure that data is collected from those groups, in proportion to the relative size of the groups?

On the research or design team, are those groups either represented or advocated for?

References:

Crawford, Kate 'The Hidden Biases in Big Data', Harvard Business Review April 1, 2013 https://hbr.org/2013/04/the-hidden-biases-in-big-data

PLAY 51: HOW WELL DO YOU KNOW YOUR CUSTOMERS?

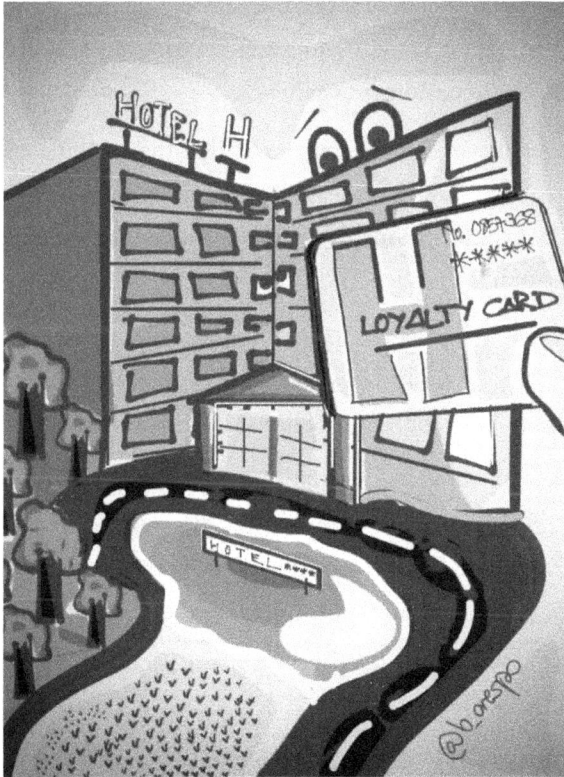

MOST HOSPITALITY LOYALTY SCHEMES ARE THE SAME, THE SAME DATA CAPTURED, THE SAME SEGMENTATIONS OF REGULAR GUESTS TO KNOW MORE ABOUT THEM. THIS PLAY DISCUSSES A DIFFERENT DEFINITION OF A LOYALTY PROGRAM.

A global hotel chain was reviewing their customer loyalty program. The program was operating successfully with millions of members accumulating and spending points. But no program, however successful, can rest on its laurels, the market is simply too competitive.

The company looked at all the attributes of its customers trying to understand what made their customers loyal. They used the lens of historical stay and spend information, preferences, and socio-demographics. This information was used to personalize the service. Collecting more personal data via guest surveys and guest tracking on and offline allowed them to improve the customer experience.

But the loyalty team also recognized that the world is changing, and consumers are increasingly aware and cautious of how their personal data is collected and used. The integrity and security of customer data is now integral to the brand promise.

Collecting more personal data and respecting privacy were opposing forces on the loyalty team: How can we intimately know our customers wants and needs by responsibly collecting and using the data we collect about them?

In this light, they asked a simple but difficult question, "Who are our most loyal customers?' To which a member of the loyalty team provided a stellar answer, 'Our most loyal customers are the ones that know us best.'

If you've ever been a regular at a particular hotel, for long term client projects or annual vacations, then after a few stays, you will start to know the hotel layout, the restaurants, the best and worst rooms, even which are the most helpful staff.

High mileage, high status air travelers know the flight times, connections, and they have preferred seats on different planes. Over their use of the service, they build a useful knowledge of your facilities or schedules.

Those customers are the most loyal irrespective of the points they accumulate.

Once you understand how to identify the customers that know you the best, then data collection turns into data sharing. Providing those customers with better information will create a reciprocal flow of their data.

Applying the Data Mindset:

Is there an activity that requires different techniques for teasing out data from your prospects or customers? Instead of working hard to capture data, what would happen if you looked for data that you can give to your customers that would help them become more effective or efficient? What could you do to create a high-volume bi-directional flow of data. How would you measure the data flow in either direction and correlate that with improved business performance.

References:

Ott, Gilbert. 'Hotel Loyalty Programs Are Playing A Dumb Game They Will Lose' GodSavethePoints August 22, 2022 https://www.godsavethepoints.com/hotel-loyalty-vs-credit-cards/

'Infographic: The 5 key drivers of customer loyalty' Loyalty Lion February 12, 2020 https://loyaltylion.com/blog/infographic-the-5-key-drivers-of-customer-loyalty

PLAY 52: TEACHING MACHINES TO BEHAVE THEMSELVES

ARTIFICIAL INTELLIGENCE LEARNS LIKE A CHILD, PRESENTED WITH BAD EXAMPLES IT WILL CREATE POOR OUTCOMES. THIS PLAY SHOWS HOW EVEN THE MOST RENOWNED TECHNOLOGY LEADERS CAN MAKE BIG MISTAKES WITH AI.

A well-respected technology leader created an A.I. chatbot that was designed to learn quickly to be able to have conversations on a microblogging platform. Launched with only a small vocabulary, it was designed to learn from the conversations it had with human users.

The chatbot had to be removed within days of its launch because it had learned to be rude and prejudiced. The design team had not accounted for the human users that it conversed with and how the conversations were instantly filled with undesirable vocabulary.

Another industry leading organization had been using A.I. systems to score job applicants to find the most suitable candidates to fill technical roles. The ground-breaking application took its learning from existing high performing employees in similar roles in the organization.

Several years in, the company realized that the system had developed a bias against female candidates. Because the employees in those roles were predominantly male where the resumes had male oriented terms, the natural language algorithms favored those words when presented with a new resume. Resumes with phrases such as "women's chess captain" or candidates from female only colleges were penalized. On this realization, the system was quickly replaced.

For any machine learning project, providing the system with the right data is vital. Systems will only learn from the data they are given and make decisions based on just that data, without judgement as to whether the outcome is able to harm any person or group.

Applying the Data Mindset:

Are you developing or commissioning an Artificial Intelligence application to solve a business problem? What are the decisions that it will make? How will the agent be trained; will it be let loose on data, or will humans be used to teach it?

What outcomes are you looking for and what outcomes do you really not want – think exhaustively about these two things. What might cause the agent to create the unwanted outcomes? How can you train the agent to avoid creating those unwanted outcomes?

References:

Lauret, Julien. 'Amazon's sexist AI recruiting tool: how did it go so wrong?' Becoming Human, August 16, 2019

https://becominghuman.ai/amazons-sexist-ai-recruiting-tool-how-did-it-go-so-wrong-e3d14816d98e

Schwartz, Oscar. 'In 2016 Microsoft's Racist Chatbot Revealed the Dangers of Online Conversation' IEEE Spectrum November 25, 2019 https://spectrum.ieee.org/in-2016-microsofts-racist-chatbot-revealed-the-dangers-of-online-conversation

A STUDY GUIDE

The book contains 52 plays, each consisting of an image, a short summary, the anecdote itself, how to apply the thinking, and finally reference texts for further reading.

The plays have been deliberately not presented in an organized fashion, so that they can be read in a random order or in sequence.

For those that require more structure or have a particular problem to resolve or a specific situation to manage, we have classified the play into one of five categories:

1.	Data Hidden in Plain Sight	The art of seeing data everywhere Plays: 2, 9, 12, 18, 23, 33, 40, 48
2.	Shifting Perspectives on Data	Re-thinking how you consider data Plays: 1, 4, 5, 8, 11, 19, 25, 30, 31, 34, 35, 37, 51
3.	The Business of Data	New and improved business models Plays: 3,6, 14, 15, 16, 20, 21, 22, 24, 28, 29, 32, 45
4.	Data Analysis Sign Flips	Alternative views of the same data Plays: 7, 10, 13, 17, 27, 36, 38, 39, 42, 44, 46, 47
5.	Responsible A.I.	Machine Learning that does no harm Plays: 26, 41, 43, 49, 50, 52

	1 Hidden	2 Perspective	3 Business	4 Analysis	5 Responsible
PLAY 1		Yes			
PLAY 2	Yes				
PLAY 3			Yes		
PLAY 4		Yes			
PLAY 5		Yes			
PLAY 6			Yes		
PLAY 7				Yes	
PLAY 8		Yes			
PLAY 9	Yes				
PLAY 10				Yes	
PLAY 11		Yes			
PLAY 12	Yes				
PLAY 13				Yes	
PLAY 14			Yes		
PLAY 15			Yes		
PLAY 16			Yes		
PLAY 17				Yes	
PLAY 18	Yes				
PLAY 19		Yes			
PLAY 20			Yes		
PLAY 21			Yes		
PLAY 22			Yes		
PLAY 23	Yes				
PLAY 24			Yes		
PLAY 25		Yes			
PLAY 26					Yes
PLAY 27				Yes	
PLAY 28			Yes		
PLAY 29			Yes		
PLAY 30		Yes			
PLAY 31		Yes			

PLAY 32			Yes		
PLAY 33	Yes				
PLAY 34		Yes			
PLAY 35		Yes			
PLAY 36				Yes	
PLAY 37		Yes			
PLAY 38				Yes	
PLAY 39				Yes	
PLAY 40	Yes				
PLAY 41					Yes
PLAY 42				Yes	
PLAY 43					Yes
PLAY 44				Yes	
PLAY 45			Yes		
PLAY 46				Yes	
PLAY 47				Yes	
PLAY 48	Yes				
PLAY 49					Yes
PLAY 50					Yes
PLAY 51		Yes			
PLAY 52					Yes

THE DATA STRATEGY WORKSHOP AGENDA

What are the objectives?

When we run data strategy workshops, we aim to do three things. Firstly, to inspire the organization and help develop a vision. Secondly to gain a better view of the data that is available, data that should be treated as a corporate asset. And thirdly, to identify and prioritize a set of execution projects that will provide a healthy return on the data asset.

Who should attend?

In person meetings and workshops are expensive, typically thousands of dollars per hour if its senior decision makers in attendance. However, to kick-start an effective data strategy, those individuals are required as they will ensure that actual projects are initiated, and their outcomes contribute to the organization's strategic vision.

In addition to executives, we ask that there should be people present who understand the business, operations, customers, and systems. Without these individuals, the picture will be incomplete and estimates less accurate. These will be the teams responsible for execution, their contribution to the strategy is critical. We recommend between 10 and 15 participants for maximum engagement and productivity.

Workshop duration

Don't be tempted to squeeze this session into a half day. Ideally this is a two-day exercise that will provide adequate time for all participants to understand, agree and disagree and to co-create an actionable plan.

Workshop Deliverables

A prioritized list of projects that could be completed in the 3 months following the workshop. The short-term deadline will favor simpler projects or initial iterations of larger projects that are more likely to demonstrate quick benefits.

During the workshop, the projects should be roughly scoped, categories of benefit identified and allocated an owner.

Agenda

This skeleton agenda shows a sequence of questions that will drive data strategy. Each day comprises 6-hour long sessions, which allows adequate time for breaks and to fully engage in conversations:

Day 1	Session Key Questions	Day 2	Session Key Questions
1 hour	Why are we here, why this organization must consider data more strategically.	1 hour	Recap of Day 1, what did we learn, what issues arose, what should we carry forward
1 hour	Introductions and individual goals, session objectives, review agenda	1 hour	What additional needs can be met using the data we have? Are there new markets for our data?
1 hour	What is data? What are the data assets that are available to us? What data would we like to have if we could?	1 hour	Are there commercial models for the raw or processed data? How can we increase the supply of data or the value we add to it?
1 hour	How can we better enable the organization with the data it needs to make good decisions?	1 hour	What potential data projects could we undertake? How can we compare and prioritize those projects?
1 hour	What broader ecosystem is our organization part of, how can data improve extra-organization activities?	1 hour	What potential risks and obstacles can we identify that might prevent the success of these projects? What mitigations are possible?
1 hour	Who are our customers? What data would we like about them? What would our customers want to share with us and why?	1 hour	What projects can we commit to taking to the next phase: feasibility, budgeting, and prototyping?

Our Experience

We have run these workshops in both academic and business contexts for numerous industry verticals. In every case we make every effort to fully engage the participants, and to ensure that every voice in the room is heard and acknowledged. The best ideas come from the least expected directions, from the contrarian, the introvert, and the subordinate. We urge you to come to these sessions with both a beginner's mind and a data mindset.

ACKNOWLEDGEMENTS

This book has been inspired by over twenty years of conversations with friends, colleagues, and clients. We have spent hours, in classrooms, at whiteboards, on calls, over meals and drinks or just walking and talking about data.

Thank you for your bright ideas and critical insights as we evolved our data mindsets, Mar Aguado, Ignasi Alcalde, Philip Andrews, David Antelo, Carola Arbolí, Luis Aribayos, Phaedra Bionodiris, Craig Brennan, David Carro, Juanjo Casado, Ramsay Chu, Darren Cornish, Andrés de Cuevas Morón, Javier de Prado, Miguel Díaz Roldán, Stephen Doran, Chris Downs, Ana Maria Echeverri, Richard Elliott, Teri Elniski, Beatriz Escriña, Adolfo Fernández Merino, Juan Luis Galán , Juan Luis Galán, Chirag Gandhi, David Gilaberte, Alfonso Javier González, Etienne Grisvard, Simon Handley, Iain Henderson, Anne Hunt, Rahel Jhirad, Chris Kalaboukis, Musaddeq Khan, To Kim, Chetan Korke, Chuck Lam, Marcelino Lominchar, Sergio López Miguel, Mike MacIntosh, Sergio Maldonado, John Marshall, John Milinovich, Lucía Miranda, Miguel Moreira da Silva, Blanca Moscoso del Prado, Carlos Muñoz, Suresh Nair, David Needham, Jim Novack, María José Pardo, Miguel Peco, Milton Pedraza, Pablo Penone Díaz, José María Pérez-Caballero Gallego, Claudia Perlich, Diego Prado, Steve Prokopiou, Ramón Puga, Shamon Ratyal, Raúl Retamosa, Damian Roca, Susana Rodríguez Urgel, Beth Rudden, David San Felipe, Jeffrey Schaubschlager, Felix Schildorfer, Lucía Schmid, Doc Searls, Ana María Seijo, Celine Takatsuno, Amit Tewari, Aparna Uberoy, Pete Warden, Andreas Weigend, Chris Wiedmann, Julian Wilson, Jason Wolfe and Arti Zeighami.

Thank you for your support, encouragement and keen eyes, Ansley Echols, Jane Gideon, Julia Link, Chris Micklethwaite, Rebecca Morton-Doherty, and Dilip Ramachandran.

Thank you to the kind and curious folks at all our clients where we were able to put our data mindsets into practice together.

And finally thank you to our students who listened, absorbed, and made these ideas their own.

ABOUT THE AUTHORS

Gam and Bernardo both live in Madrid, both teach at IE Business School and are working together on a software application that uses enterprise exhaust data to visualize and streamline business processes.

Bernardo Crespo is an entrepreneur, startup investor and also a venture builder advisor to various data-driven companies. He has also been Academic Director at IE Business School's Digital Transformation Executive Program for the last ten editions of the program. Previously he was Digital Transformation Leader at Merkle Spain and also Head of Digital Marketing at BBVA in Spain where he led a data intensive initiative based on Gamification mechanics that was a case study by prestigious technology firms such as Gartner and Forrester. He studied the last year of his undergraduate degree in Business Administration at the University of St Andrews in Scotland, graduated from UCLM in Spain as BBA, and is also a certified ontological coach by Newfield Network. Bernardo lives in Spain, where he has been recognized as one of the top 50 influencers in Digital Transformation by Expansión newspaper in 2016.

Gam Dias is a partner at UK-based Digital Transformation consultancy, 3PointsDIGITAL where he helps organizations understand and leverage data as a corporate asset. He teaches Data Strategy at IE Business School in Madrid and previously co-founded the ecommerce consultancy First Retail in Silicon Valley. He began his career in the UK as a Management Information Systems developer, he went on to become a product manager for BI and Analytics vendor Hyperion, and he managed the product and research team for a text analytics startup.

As a consultant, he has helped write the data strategy for Fortune Global 500 companies, innovative startups, and ambitious non-profits. He has a degree in Computer Science from the University of Liverpool and an MBA from Warwick Business School. Gam has lived in London, Leeds, Salt Lake City, Santa Cruz, San Francisco, and he currently lives in and works from Madrid, Spain.

www.ingramcontent.com/pod-product-compliance
Lightning Source LLC
Chambersburg PA
CBHW070519200326
41519CB00013B/2853